KU-572-829

Contents

RS•C

The role of practical work in chemistry

About practical work

'Chemistry is an experimental science and its development and application demand a high standard of experimental work' (Hanson *et al*, 1993). This, and many similar statements, can be found littering chemical and education literature. It is difficult to argue against this general thesis when it comes to the pursuit and pursuance of chemistry. Dall'alba (1993) extends the idea and asserts that an important factor in higher education teaching is to initiate students into what it is like to be a practitioner of their subject, but it is its extension to the learning of chemistry that can give rise to a divergence of views. Current thinking compasses mass education where the majority of students in chemistry may have no intention of pursuing chemistry as a career. In these circumstances it seems inappropriate to design a programme that is solely directed at training the professional chemist.

Notwithstanding this dichotomy for the teacher, it is essential that all students are able to appreciate (and experience) the constraints, potential and tensions of the investigative process. For chemistry, this necessarily involves a laboratory experience although investigative work is certainly not confined to the laboratory.

Laboratory work in chemistry is an expensive activity. Laboratories are costly to build and equip, and academic and technical staffing, instruments and consumables are a drain on resources. It is probable that restrictions imposed by safety legislation on the use and disposal of chemicals have had a major effect on practical work, particularly in the less well-endowed institutions. The perception is that it is becoming increasingly difficult to provide students with a high quality conventional practical experience. A Royal Society of Chemistry (1994) report states that 'the restrictions on resources and the time allocated to practical work are causing a decline in the extent of practical work and the standards achieved'.

An often unpopular question to ask is 'what is laboratory work for?' In other words what are the objectives, what are the outcomes? In general, the outcome is not just to train the professional scientist. It would not be reasonable to expect all students of literature to become professional writers or poets, so we must not operate on the assumption that all chemistry students will become professional scientists. Indeed in 1993, only about forty per cent of BSc chemistry majors leaving United Kingdom universities found employment or further training as professional chemists. Even where it can be argued that professional training is the focus, some practical courses do not address this aim effectively. For most students, it is a development of the ideas and approaches to science and scientific investigation that is paramount. It does not necessarily follow that an extensive experience in a well-equipped laboratory will achieve this end.

So often, it is time in the laboratory that is specified as some kind of measure

Progressive development of practical skills in chemistry
– a guide to early-undergraduate experimental work

Stuart W Bennett and Katherine O'Neale

Marjorie Cutter Scholarship

The Royal Society of Chemistry
1996–1997

LIVERPOOL
JOHN MOORES UNIVERSITY
AVRIL ROBARTS LRC
TITHEBARN STREET
LIVERPOOL L2 2ER
TEL. 0151 231 4022

LIVERPOOL JMU LIBRARY

3 1111 00929 9429

ROYAL SOCIETY OF CHEMISTRY

**Progressive development of practical skills in chemistry
– a guide to early undergraduate experimental work**

Compiled and developed by Stuart W Bennett and Katherine O'Neale, Open University
Edited by Denise Rafferty and Sara Sleigh
Designed by Imogen Bertin and Sara Roberts
Published by The Royal Society of Chemistry
Printed by The Royal Society of Chemistry

Copyright © The Royal Society of Chemistry 1999

Apart from any fair dealing for the purposes of research or private study, or criticism or review, as permitted under the UK Copyright Designs and Patents Acts, 1988, this publication may not be reproduced, stored, or transmitted, in any form or by any means, without the prior permission in writing of the publishers, or in the case of reprographic reproduction, only in accordance with the terms of the licences issued by the Copyright Licensing Agency in the UK, or in accordance with the terms of the licences issued by the appropriate Reproduction Rights Organization outside the UK. Enquiries concerning reproduction outside the terms stated here should be sent to the Royal Society of Chemistry at the London address printed on this page.

For further information on other educational activities undertaken by the Royal Society of Chemistry write to:

The Education Department
Royal Society of Chemistry
Burlington House
Piccadilly
London W1V 0BN

Email: education@rsc.org

ISBN 0–85404–950–9

British Library Cataloguing in Data.

A catalogue for this book is available from the British Library.

of quality of experience rather than an assessment of skills. Time spent in laboratories is not always well used. In a recent survey, based on work by Maskill and Meester (1993), we have found that, on average, a student in the first year of chemistry courses in English universities performs over fifty titrations. Although these operations form part of a range of experiments, it cannot be a valuable learning experience to carry out extensive repetition of this relatively simple manipulation. Another problem relates to active participation in experiment design. How often does the material supplied to students read like a recipe and how often is it treated like a recipe by the student? The questions to ask are 'what skills should be developed in students, which of these skills are traditionally developed in the laboratory, and can any of these be developed outside the expensive laboratory environment?' It is impossible to home in on a unique list of skills, but the major skills that are ideally developed in a laboratory environment include:

- manipulation;

- observation;

- data collection;

- processing data;

- analysing data and observations;

- interpretation;

- problem solving;

- team work;

- experiment design;

- communication and presentation; and

- laboratory ambience.

A major problem faced by students is in the progressive development of these skills as they move through an undergraduate course. Each laboratory experience may be valuable and worthy in its own right. However, the next session (or even the next semester) in the laboratory may not take into account the extent of skills developed in the earlier session. Indeed, even today it is not common to find laboratory programmes analysed, let alone designed, in a skills development context. To move in the direction of a skills driven programme is not only central to the quality of student progress but results in a more efficient use of the laboratory resource.

Most teachers of chemistry have been faced in the laboratory with such questions as 'Is this right?' while a student proffers a white powder. The reaction to the enquiry 'What is it?' is often 'Well it's, er, here' as the student points to part of their laboratory handbook. Exchanges such as this are, sadly, often interpreted as the fault of the student failing to 'read ahead'. It just might not be entirely the fault of the student. The parallel argument is that if a train is consistently late by twenty minutes each day then it might be that the timetable is inappropriate. So, perhaps there is something seriously awry with the way that practical work is often designed.

Too often, time spent in a laboratory is regarded as an indicator of competence level in practical work, despite the fact that practical work transcends the laboratory to other activities. Criteria for course approval sometimes specify a minimum number of hours in the laboratory. In the past, too little effort has been put into a focus on what the laboratory is good at and where it is weak. Possibly the advent of electronic media (and, in particular, the CD-ROM) has helped bring skills development to the fore and encouraged a consideration of which practical skills can be developed (at least to some extent) outside the laboratory. As with all learning experiences, an evaluation of outcomes is a more useful parameter than hours spent.

The motivation for this work came from a belief that some undergraduate students are not getting good value from the long hours that are traditionally spent in the laboratory, and that they often develop a negative attitude to the experience. It can be argued that the 'recipe' style presentation of laboratory practicals (I hesitate to use the term experiments) is a major contributor. However, the problem is in the way that 'recipes' are used. Often the student reads through the notes line by line, mechanically carrying out the manipulations, with no real thought as to why certain actions are taken and how they fit into the overall outcome. The parallel with research is useful here. Much of the intellectual effort comes before entering the laboratory: discussions, literature work, design of the experiment, sorting out quantities, conditions and equipment. This level of pre-laboratory activity is often denied the undergraduate and yet it forms a vital, essential and stimulating component.

Good teaching and good learning are contingent on an appropriate environment. Anyone faced with too many unfamiliar things to do at one time is bound to perform poorly. Learning to drive a vehicle is a good example. The first outing on the road confronts the learner with the need to steer, brake or accelerate, change gear and observe (over 360°), and to carry out several of these activities at the same time. Even changing gear is a complex multi-part operation of split-second timing (release accelerator, depress clutch, move selector appropriately, adjust accelerator and smoothly engage clutch) whilst making observations and steering the vehicle. It is not possible to drive along a road and concentrate on gear changing to the exclusion of everything else. So how is it that experienced drivers are able to drive safely and listen to the radio at the same time? The answer lies in the concept of 'working space' expounded most eloquently by Johnstone (1997) in his Brasted Lecture address. There comes a point when it is not possible to process or work with more than a certain number of pieces of information (usually six to eight) at the same time. Experienced practitioners overcome this limitation by gathering all the steps in gear changing under one activity, a process known as 'chunking'. In this way, changing gear is a single activity. Much the same process enables the representation of nitrobenzene to be immediately recognisable to a chemist as comprising two components, a phenyl group and a nitro group. To the initiate, the picture is of thirteen lines and three symbols (two alphabetic and one numerical).

Johnstone's analysis has had significant impact on the learning of chemistry but it is arguably in the laboratory where heed to these lessons is in greatest need. The student (particularly the fresher) enters the laboratory 'cold' except for perhaps a short discourse on safety rules. The inputs of information are huge, for example, location of chemicals and identification of the particular materials needed to begin the prescribed work, recognition of equipment and its handling, instrumentation, and safety requirements. It should not be surprising that most students are unable to give much intellectual effort to the structure of the laboratory activity. Indeed, the ability to plough through a 'recipe experiment' line by line could be regarded as a major achievement in such circumstances.

There are (at least) two complementary approaches that should help alleviate this unsatisfactory situation. The first is very simple and is addressed towards easing the student's familiarity with unfamiliar surroundings. Provide each student with simple drawings or photographs of the equipment that is needed for a laboratory session. It is simply not fair to expect a new student to spot a Hirsch funnel at twenty paces! A plan of the laboratory with the location of all the chemicals and equipment that are needed for the session would help avoid many of the 'Where is it?' questions that typically occupy much of a demonstrator's time in the early part of a laboratory session. The second approach is more fundamental and it is an approach upon which this book is based.

In an attempt to limit the demands on the student (and the poor learning environment that ensues if demands are unreasonable), we have gathered (and developed) twenty-one practical activities which span a wide range of chemistry. These tried and tested activities are intended to be representative and not comprehensive and are directed toward the early part of an undergraduate chemistry programme. Each activity has been analysed from a skills, rather than from a content, standpoint and the activities ordered in increasing skills demand (both in level and sophistication). This approach takes due regard of the entry behaviour of the student and acts as a focus for defined outcomes.

The series of activities is not intended to be prescriptive of a programme and it would be entirely appropriate to select individual activities that could be slotted into an existing course. However, the series does illustrate that it is possible to develop a programme that ranges over chemistry and results in students developing a pre-defined portfolio of skills. Your skills outcome requirements will almost certainly be different from this and will necessitate a different programme. Nevertheless, by starting with desired outcomes and selecting and developing activities that collectively achieve these, there is much less risk of omitting the development of specific skills and also of unnecessary over-emphasis of particular skills.

Stuart W Bennett

How to use this book

The activities in this book are arranged in order of increasing skills development and demand. Each activity includes a Student Guide (**denoted by S**) containing the material from which the student works. These notes have been written in several styles, one of which merits singular mention. Many students find it difficult to maintain familiarity with the logic and the structure of a piece of work while, at the same time, working on a small detail. Presenting students with a breakdown of the activity in terms of a flow chart has been found to be particularly useful. For each box on the chart (which contains the basics of that step) there is a corresponding sheet indicating details of equipment, chemicals, process and safety. Inexperienced students seem to be able to focus better with this format, although it is only suitable for particular activity structures.

The Student Guide is accompanied by a Demonstrator Guide (**denoted by D**) aimed at teachers. These notes include specific comments on the processes and comments on the questions included in the Student Guide. Ideas for pre-laboratory activities are also given. It cannot be emphasised just how important the pre-laboratory activity is. Pre-laboratory activities are more than simply telling students that they should read through the notes before the next session. They should be student active. As mentioned earlier, there is a place for the 'recipe' type activity (and it can even parallel research) provided the active pre-laboratory session involves problem identification, solution strategy and experiment design. Another area which is often neglected is the post-laboratory session, which is always valuable and is essential for those activities that involve a team approach with individual members working on different aspects of a problem. Although not included in this publication, great attention should be paid to assessment and evaluation of practical work in the context of the defined objectives. The detailed review by Laws (1996) of research into undergraduate science education contains a substantial section on laboratory work.

As an aid to setting up the laboratory, each activity includes a Technical Guide (**denoted by T**) which details all reagent and equipment requirements.

Skills analysis

Skills analysis of any activity is not simple. There is a hierarchy of skills based on intellectual and manipulative demands and even then, specific skills require careful definition. The only effective way of defining a skill is by detailing exactly what the student is able to do once that skill has been acquired. An outcome that states 'interpret an infrared spectrum' is not sufficient. What kind of spectrum: gas or liquid (or mull), absorption or reflectance, what frequency range, group frequencies or normal modes? To do the job effectively, outcomes have to be written in terms of the behavioural objective.

Even a superficial attempt to analyse activities for skills can lead to useful indicators. Identification of over- and under-emphasis of particular skills and subsequent fine-tuning of the practical programme can result in an increased efficiency in the use of the laboratory. It is also worth asking whether some of the skills can be developed (at least in part) outside the laboratory. Experiment design can be seen in this context and it would not be unreasonable to suggest that all of the skills in the earlier list (with the exception of manipulation and laboratory ambience) can be developed to some extent outside the laboratory. Continuing improvements in the quality of multimedia software make this line of approach increasingly fruitful.

Each activity in this book has been allocated to one of the four general categories proposed by Kirschner and Meester (1988).

A   the academic or formal laboratory which employs didactic methods to verify and illustrate laws and concepts – ie **formal** (verify concepts);

B   the experimental laboratory in which exercises are open-ended and relatively unstructured – ie **experimental** (relatively unstructured);

C   the divergent laboratory which offers tasks with an initial, standard, structured component which may be developed in a number of different ways – ie **divergent** (variable development from common start);

D   the investigatory skills-teaching laboratory in which the procedures of investigation are the principal subjects of study – ie **investigatory** skills.

By changing the style of notes and the information supplied to the students, the category of some activities will change.

The skills analysis system has been distilled onto a single form. The major skills categories have been subdivided into their different facets. For example, team work can include problem identification and analysis (skills which are also well placed under problem solving). Team work skills also include identification of the personal skills required to solve a problem, selection of the team on the basis of the members' strengths (and weaknesses), development of a strategy, assignment of roles to team members based on individual strengths, organisation of the team operation (timescales, reporting, redirection), evaluation and optimisation of resources and development and communication of outcomes. Each of these categories can be further sub-analysed and the same is true for all of the skills categories. We have tried to limit the analysis of each activity to a level that is quick and easy to carry out, yet carries sufficient information to provide a useful input to a collective course skills package.

The analysis form is not unique and an amended version may be more useful for your particular programme. A quick and simple way of using the form is to reproduce it on a transparent sheet. By overlaying the completed sheets for the practical programme it is easy to see which categories may be over-developed and which neglected. A summary form, on which the skills for a complete experimental programme are listed, can also provide a useful first attempt at identifying over- and under-developed skills. A more definitive and detailed analysis can be done electronically.

Activity type

formal ○	experimental ○	divergent ○	investigatory ○

Skills

manipulation	weighing ○	volume ○	handling ○			
techniques	reflux/distillation ○	recrystallisation ○	chromatography ○	inert atmosphere ○	spectroscopy ○	titration ○
observation	colour ○	volume ○	temperature ○	pressure ○	physical state ○	
data collection	qualitative ○	numerical ○	spectral ○	electronic ○		
data processing	calculation ○	computing ○	matching ○			
interpretation	selection ○	validation ○	deduction ○	prediction ○		
problem solving	identification ○	in/output ○	breakdown ○	methods ○	assembly ○	
team work	skills identification ○	analysis ○	role assignment ○	organisation ○	resources ○	outcomes ○
experiment design	input ○	output ○	precision ○	techniques ○	validity ○	
communication	report ○	poster ○	oral ○	audience ○		
safety	COSHH ○	application ○	review ○	disposal ○		

Index of activities

Activities are grouped according to the four general categories outlined previously:

Type A – the academic or formal laboratory which employs didactic methods to verify and illustrate laws and concepts.

Type B – the experimental laboratory in which exercises are open-ended and relatively unstructured.

Type C – the divergent laboratory which offers tasks with an initial, standard, structured component which may be developed in a number of different ways.

7 **Properties of a zeolite:** Group activity to identify the properties of a zeolite extracted from detergent washing powder.

14 **Synthesis of a dye stuff intermediate and an azo dye:** Two different routes to the same end-product are investigated and compared in terms of yields and costs.

18 **Electrochemical cells:** Team work is involved here to build up a picture of how electrochemical cells operate. Concentration dependence of cell electromotive force and measurement of electrode potential provide the focus.

Type D – the investigatory skills-teaching laboratory in which the procedures of investigation are the principal subjects of study.

3 **Aspirin analysis:** A quantitative investigation to estimate the proportion of active ingredient in commercial aspirin tablets.

8 **Nicotine and smoking:** Techniques including extraction methods and gas chromatography are developed in this activity which is based on the extraction of nicotine from a range of tobaccos.

10 **Synthesis and analysis of an aluminium oxalate complex:** This investigation develops skills in synthesis and volumetric analysis.

11 **Catalase investigation:** This activity involves an investigation into the factors which influence the rate of decomposition of hydrogen peroxide with catalase. The work can be carried out either individually or in groups.

Note that references to first and second year students relate to the English university system.

Acknowledgements

We are deeply indebted to so many for help with this project.

Financial support in the form of a Marjorie Cutter Scholarship from the Royal Society of Chemistry is gratefully acknowledged. The project has been administered by Dr Denise Rafferty, Education Officer, Professional Practice, the Royal Society of Chemistry, with whom we have had many valuable discussions.

We wish also to thank the Open University for providing facilities and access to its practical programmes which have formed the basis for a number of activities. We also value the many discussions with colleagues at the Open University.

In addition we extend our sincere thanks to the following who have provided materials and offered constructive advice.

Professor P N Bartlett, University of Southampton
Dr W J Bland, Kingston University
Dr S W Breuer, Lancaster University
Dr R H Dahm, De Montfort University
Dr M Douglas, Sheffield Hallam University
Dr M J Frearson, University of Hertfordshire
Dr C J Garratt, University of York
Dr M Greenhall, University of Huddersfield
Dr T K Halstead, University of York
Dr J R Hanson, University of Sussex
Dr C A Heaton, Liverpool John Moores University
Dr R A Hill, Glasgow University
Dr J A Hriljac, University of Birmingham
Dr D Ingram, Aston University
Dr J C John, University of Dundee
Dr J H Keeler, University of Cambridge
Dr A Kumar, GlaxoWellcome plc
Dr R Lowry, University of Plymouth
Dr J McGinnis, University of Teesside
Dr R Maskill, University of East Anglia
Dr C Mortimer, University of Central Lancashire
Dr B P Murphy, Manchester Metropolitan University
Dr D C Nonhebel, University of Strathclyde
Dr T L Overton, University of Hull
Dr C Peacock, Lancaster University
Professor J Pearce, Liverpool John Moores University
Dr S D Price, University College London
Dr A J Rest, University of Southampton
Dr A Russell, University of Newcastle upon Tyne
Dr R Simmonds, University of Wales, Aberystwyth
Dr M Smith, University of Leeds
Dr A Vincent, Kingston University
Dr J L Wardell, University of Aberdeen
Dr B J Walker, Queen's University of Belfast

References

1   J R Hanson, J Hoppé and W H Pritchard, *Chem. Br.*, 1993, **29**, 871.

2   G Dall'alba, *Learning and Instruction*, 1993, **3**, 299.

3   *The design and delivery of degree courses in chemistry leading to professional membership*, 1994, The Royal Society of Chemistry.

4   M A M Meester and R Maskill, *First year practical classes in undergraduate chemistry courses in England and Wales*, 1993, The Royal Society of Chemistry.

5   A H Johnstone, *J. Chem. Educ.*, 1997, **74**, 262.

6   P M Laws, *Studies in Science Education*, 1996, **28**, 1.

7   P Kirschner and M A M Meester, *Higher Education*, 1988, **17**, 81.

Bibliography

PROGRESSIVE
DEVELOPMENT
OF PRACTICAL
SKILLS IN
CHEMISTRY

■

xiii

The following practical texts may be useful to use in conjunction with this book.

1. W R Moore and A Winston, *Experiments in Organic Chemistry*, McGraw-Hill International Publishers, 1996.

2. A M Halpern and J H Reeves, *Experimental Physical Chemistry: A Laboratory Approach*, Prentice-Hall, 1997.

3. G P Matthews, *Experimental Physical Chemistry*, Oxford University Press, Oxford, 1986.

4. M Singh, *Microscale and Selected Macroscale Experiments for General and Advanced General Chemistry*, John Wiley & Sons Inc., New York, 1994.

5. R J Errington, *Advanced Practical Inorganic and Metalorganic Chemistry*, Blackie, 1997.

1. Hydrogen rocket

This activity is suitable for an introduction to the chemistry laboratory. The demands made on students in terms of manipulative skills are minimal, however it provides the opportunity to develop skills of scientific investigation.

Students use 'rockets' powered by a mixture of hydrogen and oxygen gas to investigate the criteria that are important in maximising the rocket's flight paths. This is an enjoyable way to practise systematic scientific investigation.

This experiment is suitable for:

■ first year students

■ two hours plus a pre-laboratory session

■ group work

Activity type

formal ○	experimental ●	divergent ○	investigatory ○

Skills

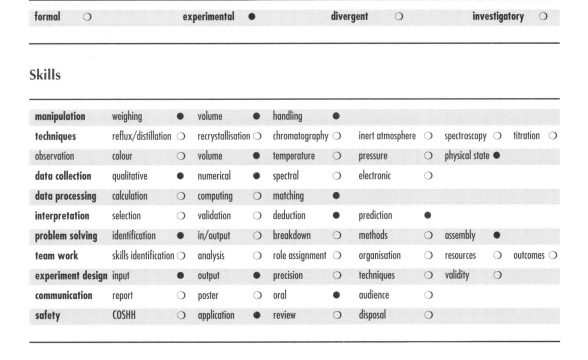

manipulation	weighing ●	volume ●	handling ●			
techniques	reflux/distillation ○	recrystallisation ○	chromatography ○	inert atmosphere ○	spectroscopy ○	titration ○
observation	colour ○	volume ●	temperature ○	pressure ○	physical state ●	
data collection	qualitative ●	numerical ●	spectral ○	electronic ○		
data processing	calculation ○	computing ○	matching ●			
interpretation	selection ○	validation ○	deduction ●	prediction ●		
problem solving	identification ●	in/output ○	breakdown ○	methods ○	assembly ●	
team work	skills identification ○	analysis ○	role assignment ○	organisation ○	resources ○	outcomes ○
experiment design	input ●	output ●	precision ○	techniques ○	validity ○	
communication	report ○	poster ○	oral ●	audience ○		
safety	COSHH ○	application ●	review ○	disposal ○		

S1

Hydrogen rocket

The aim of this activity is to investigate the factors that are important in maximising the distance travelled by a 'rocket' fuelled by a hydrogen-oxygen gas mixture.

Equipment

- 3 x 5 cm^3 plastic pipettes
- spatula
- 2 x boiling-tubes
- 2 x rubber bungs (bored with 2 mm nylon tubing)
- 2 x 100 cm^3 beakers
- large rubber bung
- paper clips
- measuring tape
- Tesla coil

Reagents

- manganese(IV) oxide, powdered
- hydrogen peroxide solution (3%)
- zinc, granulated
- hydrochloric acid solution (4.0 mol dm^{-3})

Safety

Note that mixtures of hydrogen gas and air are explosive. Keep naked flames away from the working area. Consult a demonstrator before using the Tesla coil.

Manganese oxide: Harmful if swallowed.

Hydrochloric acid solution (4.0 mol dm^{-3}): Causes burns. Irritating to eyes, respiratory system and skin.

Hydrogen peroxide (3%): Irritant to skin, eyes and mucous membranes.

Generating oxygen

Fill a plastic pipette completely with water. Place a spatula-end of powdered manganese(IV) oxide into a boiling-tube and carefully add 3% hydrogen peroxide solution to a depth of about 4 cm. Stopper the tube with the bored rubber bung fitted with a length of nylon tubing, and clamp it vertically. Insert the end of the tubing into the water-filled pipette, and clamp the pipette vertically, with the bulb at the top. Use a large beaker to collect the water dripping from the pipette. Remove the tubing when the pipette is almost full of oxygen, leaving a small plug of water to seal the end. Test the gas with a glowing match to confirm that oxygen has been produced (consult a demonstrator if this procedure is unfamiliar).

Write a chemical equation to represent the reaction.

Generating hydrogen

Fill a plastic pipette completely with water. Place a spatula-end of powdered zinc into a boiling-tube and add 4.0 mol dm^{-3} hydrochloric acid to a depth of about 4 cm. Collect the hydrogen in the same manner that oxygen gas was collected in the last section. Use a flame to confirm that hydrogen has been produced (consult a demonstrator if this procedure is unfamiliar).

Write a chemical equation to represent the reaction.

Preparing a rocket

Renew the reagents in the gas generators and use them to fill a plastic pipette with approximately equal volumes of oxygen and hydrogen. Leave a small plug of water in the bottom of the 'rocket' to act as a seal.

Use the large rubber bung and a paper clip to fashion a launch pad for the rocket. Ensure that the launch path is clear. Insert the straightened end of a paper clip into the pipette past the plug of water, then launch the rocket by igniting the hydrogen/oxygen mixture with the Tesla coil. Measure the length of the flight.

Further investigation

How can the distance the rocket flies be increased? Make a list of the factors that might be influential and devise ways of evaluating their effects.

Are there other ways of generating hydrogen and oxygen? Can they be generated from a single process? After consultation with a demonstrator, design experiments to try out any of these ideas.

Hydrogen rocket

This is a relatively simple and enjoyable activity, which could act as a student introduction to the laboratory. A two hour session is all that is required. The main aim is for students to develop problem solving skills, and to gain confidence in working in a laboratory. Students have been given directions for generating hydrogen and oxygen, and basic instructions in building a rocket, but are deliberately not given explicit instructions. It is important that students are given clear directions in a pre-laboratory session and are encouraged to talk through their procedure with colleagues.

The equations for the reactions are as follows:

$$MnO_2 + 2H_2O_2 \rightarrow Mn + 2O_2 + H_2O$$

$$Zn + 2HCl \rightarrow ZnCl_2 + H_2$$

The main problem in this activity is that, if there is undue delay in connecting the gas generators to the pipettes, there may be insufficient gas remaining to complete the filling. Additional reagents can be added without cleaning the used generators.

Aluminium and sodium hydroxide solution ($5.0 \ mol \ dm^{-3}$) can be used as an alternative source of hydrogen, and post-laboratory discussion could include the amphoteric properties of some metals. Electrolysis of (acidified) water is a source of both hydrogen and oxygen (conveniently in a 2:1 volume ratio). This can be achieved using a 9V battery (MN) as a simple power supply, which is connected to carbon electrodes (pencil 'leads') in a U-tube containing acidified water.

References

1   For information about the hydrogen rocket activity:
    G. Wollaston, *J. Chem. Educ.*, 1995, **72**, 1128.

2   For information about the electrolysis of water:
    C. Suzuki, *J. Chem. Educ.*, 1995, **72**, 912.

Hydrogen rocket

Equipment

- 3 x 5 cm^3 plastic pipettes
- spatula
- 2 x boiling-tubes
- 2 x rubber bungs (bored with 2 mm nylon tubing)
- 2 x 100 cm^3 beakers
- large rubber bung
- paper clips
- measuring tape*
- Tesla coil*

*One measuring tape and one Tesla coil can be shared by up to twelve students.

Reagents

- manganese(IV) oxide, powdered
- hydrogen peroxide solution (3%)
- zinc, granulated
- hydrochloric acid solution (4.0 mol dm^{-3})
- aluminium, powdered
- sodium hydroxide solution (5.0 mol dm^{-3})

LIVERPOOL
JOHN MOORES UNIVERSITY
AVRIL ROBARTS LRC
TITHE...
LIVER...
TEL...

2. Aspirin synthesis

2-Ethanoyloxybenzoic acid (commonly known as aspirin) is synthesised by reacting 2-hydroxybenzoic acid with ethanoic anhydride. Students are shown each technique by a demonstrator before they carry it out themselves. This introductory activity represents a three hour laboratory session.

Ideas for further investigation are given in **D2**, (p. 9) some of which may be suitable for more advanced students. The activity can be combined with Aspirin Analysis (p. 13).

This experiment is suitable for:

■ first year students

■ approximately three hours

■ individuals/pairs

Activity type

formal	●	experimental	○	divergent	○	investigatory	○

Skills

manipulation	weighing	●	volume	●	handling	●				
techniques	reflux/distillation	●	recrystallisation	●	chromatography	○	inert atmosphere	○	spectroscopy ●	titration ○
observation	colour	○	volume	○	temperature	●	pressure	○	physical state ●	
data collection	qualitative	●	numerical	●	spectral	●	electronic	○		
data processing	calculation	○	computing	○	matching	●				
interpretation	selection	○	validation	●	deduction	○	prediction	○		
problem solving	identification	●	in/output	○	breakdown	○	methods	○	assembly ●	
team work	skills identification	○	analysis	○	role assignment	○	organisation	○	resources ○	outcomes ○
experiment design	input	○	output	○	precision	○	techniques	○	validity ○	
communication	report	●	poster	○	oral	○	audience	○		
safety	COSHH	○	application	●	review	○	disposal	○		

S2

Aspirin synthesis

The aim of this activity is to synthesise and purify 2-ethanoyloxybenzoic acid (acetylsalicylic acid, more commonly known as aspirin) and confirm the identity of the product by its melting point and infrared spectrum.

Equipment

- 100 cm^3 round-bottomed flask, one neck
- 25 cm^3 measuring cylinder
- water-cooled condenser
- 500 cm^3 beaker
- 25 cm^3 beaker
- weighing bottle and lid
- spatula
- rubber tubing
- heating mantle
- Buchner funnel
- Buchner flask
- filter paper
- glass stirring rod
- anti-bumping granules
- melting point apparatus
- top pan balance

Reagents

- ethanoic acid (100%)
- ethanoic anhydride
- 2-hydroxybenzoic acid

Safety

Ethanoic acid and ethanoic anhydride: Flammable. Corrosive. Irritating vapour.

2-Hydroxybenzoic acid: Harmful on contact with skin.

Ethanoic anhydride should be kept in the fume cupboard. Gloves should be worn whilst handling 2-hydroxybenzoic acid, ethanoic anhydride and ethanoic acid.

Synthesis

A demonstration of each part of the synthesis will be given. Sections 1 and 2 must be carried out in a fume cupboard.

1   Mix 10 cm^3 of ethanoic acid with 10 cm^3 of ethanoic anhydride in a 100 cm^3 round-bottomed flask. Add 10 g of 2-hydroxybenzoic acid and a few anti-bumping granules and attach a water-cooled condenser to the flask. Reflux the mixture for 30 minutes.

2   Pour the mixture slowly, with vigorous stirring, into 200 cm^3 of cold water. Remove the precipitated 2-ethanoyloxybenzoic acid by filtration through a Buchner funnel.

3   Recrystallise the sample from 50% (volume) ethanoic acid/water. Dry the product on the filter paper and then transfer it to an oven (80 °C) for further drying. Weigh the dried product, and record its infrared spectrum and melting point.

Post-laboratory work

Look up the literature value for the melting point of 2-ethanoyloxybenzoic acid. How does it compare with the value obtained for the product?

Compare the infrared spectrum taken with that of an authentic sample of 2-ethanoyloxybenzoic acid and identify the main absorbances.

Write a chemical equation to represent the reaction. 2-Hydroxybenzoic acid is the limiting reagent in this reaction. Calculate the percentage yield of product.

Aspirin synthesis

The aim of this activity is to synthesise and purify 2-ethanoyloxybenzoic acid (commonly known as aspirin) and confirm the identity of the product by its melting point and infrared spectrum.

This activity could represent an introduction to the organic laboratory but, in terms of skills development, it follows reasonably from the introductory activity (p. 1). The 2-ethanoyloxybenzoic acid product of this activity can be introduced into the following activity, in which aspirin tablets are analysed.

Synthesis

The first two parts of the procedure described in the student guide should be carried out in a fume cupboard. Students should be shown each part of the procedure at the appropriate time, before they are allowed to carry it out. First they should be shown how to set up the reaction for reflux. While the 2-ethanoyloxybenzoic acid is precipitating, vacuum filtration with a Buchner funnel should be demonstrated, as should recrystallisation. The melting point of the product can be recorded by students both before and after recrystallisation if necessary.

Post-laboratory work

The post-laboratory work for this experiment is limited so that students may concentrate on the manipulative aspects of their work. The questions focus on experimental procedure as much as on background theory. Students will probably need to be told where they can obtain a literature value for the melting point of 2-ethanoyloxybenzoic acid (138–140 °C), and how to identify the main absorbances in its infrared spectrum.

Infrared spectrum of (2-ethanoyloxybenzoic acid)

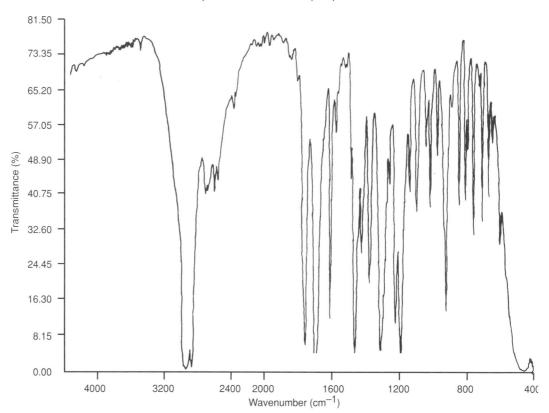

If x g is the mass of 2-hydroxybenzoic acid used (and it is regarded as the limiting reagent), the yield is given by

$$\frac{y \times 138.1}{x \times 180.2}$$

where y is the mass of product (g).

Further investigation

There are a number of variations of this activity that would enable it to be the focus of group work.

Different reaction routes
The following reaction has been carried out.

Concentrated sulfuric acid can be used in the place of ethanoic acid in the reaction. Students could investigate changing the catalyst: can the reaction be base-catalysed?

2-Ethanoyloxybenzoic acid can be synthesised by different routes, but none are as straightforward as acetylating the hydroxyl group on 2-hydroxybenzoic acid. Most other reaction routes are multi-step syntheses. It can be valuable to introduce students to some limited literature work here, by asking them to compile a list of alternative synthetic routes. These could be discussed in the post-laboratory session and some even used in a future laboratory session.

Different ratios/amounts

Students need to complete a simple equation to calculate the quantities of 2-hydroxybenzoic acid and ethanoic anhydride needed. However, the quantity of acid needed is not so clear. Students could investigate the relative reaction rates with different concentrations of acid. There is a point where the addition of further acid does not increase the reaction rate.

Different reflux times

Students could investigate the effect of changing the reflux time by following the progress of the reaction by thin layer chromatography or by precipitating 2-ethanoyloxybenzoic acid from aliquots of the reaction mixture. The second method requires some skill and careful work in crystallisation with minimal product loss.

Different solvents for crystallisation

This would be a good opportunity to discuss crystallisation. Students could investigate which solvent is effective by dividing their crude product into portions and recrystallising each portion with a different solvent. Alternatively a group discussion could be used to explore different solvent choices.

Aspirin synthesis

Equipment

- 100 cm^3 round-bottomed flask, one neck
- 25 cm^3 measuring cylinder
- water-cooled condenser
- 500 cm^3 beaker
- 25 cm^3 beaker
- weighing bottle and lid
- spatula
- rubber tubing
- heating mantle
- Buchner funnel
- Buchner flask
- filter paper
- glass stirring rod
- anti-bumping granules
- melting point apparatus
- top pan balance

Reagents

- ethanoic acid (100%)
- ethanoic anhydride
- 2-hydroxybenzoic acid

3. Aspirin analysis

In this experiment students are introduced to quantitative work. Aspirin tablets, which are made from 2-ethanoyloxybenzoic acid, are hydrolysed by boiling with an excess of sodium hydroxide solution to give 2-hydroxybenzoate and ethanoate anions. The remaining sodium hydroxide is titrated against standardised hydrochloric acid and the amount of 2-ethanoyloxybenzoic acid originally present is calculated. The information obtained is compared with the analysis found on a packet of aspirin tablets. The whole activity, including calculations, can be carried out in four hours.

This activity is presented in two different styles. The first is a conventional procedure, which demands a good pre-laboratory session, whereas the second uses a flow chart in which individual steps are expanded. These indicate how a change of format can improve an experiment.

This activity has been adapted from a practical activity used at Kingston University.

This experiment is suitable for:

■ first year students

■ approximately four hours

■ individuals/pairs

Activity type

formal	○	experimental	○	divergent	○	investigatory	●

Skills

manipulation	weighing	●	volume	●	handling	●						
techniques	reflux/distillation	○	recrystallisation	○	chromatography	○	inert atmosphere	○	spectroscopy	○	titration	●
observation	colour	●	volume	●	temperature	○	pressure	○	physical state	○		
data collection	qualitative	○	numerical	●	spectral	○	electronic	○				
data processing	calculation	●	computing	○	matching	○						
interpretation	selection	○	validation	○	deduction	○	prediction	○				
problem solving	identification	○	in/output	○	breakdown	○	methods	○	assembly	○		
team work	skills identification	○	analysis	○	role assignment	○	organisation	○	resources	○	outcomes	○
experiment design	input	○	output	○	precision	○	techniques	○	validity	○		
communication	report	●	poster	○	oral	○	audience	○				
safety	COSHH	○	application	●	review	○	disposal	○				

Aspirin analysis

The aim of this experiment is to determine the purity of commercially available aspirin tablets. The procedure has three main stages – standardisation of sodium hydroxide solution; processing the aspirin tablets; and titrations to determine the purity of the tablets. The potassium hydrogen phthalate solution used in the standardisation of sodium hydroxide must also be prepared.

Equipment

- 2 x weighing bottles
- 2 x 250 cm^3 volumetric flasks
- funnel
- 10 cm^3 pipette
- 20 cm^3 pipette
- 50 cm^3 pipette
- pipette filler
- 50 cm^3 burette
- 2 x 100 cm^3 conical flasks
- 2 x 50 cm^3 conical flasks
- 250 cm^3 beaker
- anti-bumping granules
- glass rod
- pestle and mortar
- spatula
- plastic pipette
- analytical balance
- top pan balance
- hot plate
- 2 x watch glasses

Reagents

- sodium hydroxide pellets
- potassium hydrogen phthalate
- phenolphthalein indicator (1% solution in ethanol)
- 10 aspirin tablets
- hydrochloric acid (0.50 mol dm^{-3})

Safety

Sodium hydroxide (solid and solution): Corrosive. Causes severe burns. Risk of permanent damage to eyes.

Phenolphthalein (1% in ethanol): Highly flammable. Toxic by inhalation, in contact with skin and if swallowed. Irritating to eyes, skin and respiratory system.

Gloves must be worn whilst handling sodium hydroxide.

Standardisation of sodium hydroxide solution

Record all the data collected in tables similar to those found below.

Weigh accurately about 12.5 g of potassium hydrogen phthalate (KHP) and transfer it to a clean beaker. Dissolve the solid in about 50 cm^3 of distilled water and transfer the solution carefully to a 250 cm^3 volumetric flask by pouring the solution down a glass rod. Rinse the beaker with several (10 cm^3) aliquots of distilled water and transfer these to the flask. Add water up to the mark (using a plastic pipette for the last few cm^3), stopper the flask and mix the solution well.

Weigh about 5 g of solid NaOH into a second weighing bottle. Transfer it to a beaker, dissolve it in about 50 cm^3 of water and pour the solution into a 250 cm^3 volumetric flask. Add water up to the mark, stopper the flask and mix the solution well.

Pipette 20 cm^3 of this sodium hydroxide solution into a 100 cm^3 conical flask and titrate it with the standard potassium hydrogen phthalate solution using phenolphthalein as an indicator. Repeat the titration twice to get titres that differ by no more than 0.2 cm^3.

The reaction between sodium hydroxide and potassium hydrogen phthalate can be represented by

$$HCO_2C_6H_4CO_2^-{}_{(aq)} + OH^-{}_{(aq)} \rightleftharpoons {}^-CO_2C_6H_4CO_2^-{}_{(aq)} + H_2O$$

One mole of the hydroxide anion reacts with one mole of potassium hydrogen phthalate.

Preparation of aspirin tablets

Weigh accurately ten aspirin tablets and determine the average mass of a single tablet. Grind four aspirin tablets using a pestle and mortar and place an accurately weighed sample of about 0.8 g of this in a beaker. Use a pipette to add 50 cm^3 of standard sodium hydroxide solution to the beaker and add a further 50 cm^3 of sodium hydroxide solution to a second beaker (to act as a control). Add a couple of anti-bumping granules to each beaker, cover them both with watch glasses and boil the contents gently for about 10 minutes. Allow the beakers to cool; rinse the undersides of the watch glasses into the beakers with distilled water.

Aspirin is the common name for 2-ethanoyloxybenzoic acid. When the 2-ethanoyloxybenzoic acid is boiled with sodium hydroxide, hydrolysis occurs and 2-hydroxybenzoic acid, ethanoic acid and water are formed.

RS•C

One mole of 2-ethanoyloxybenzoic acid reacts with two moles of hydroxide ions.

Determination of purity of aspirin tablets

Using a pipette, put a 10 cm^3 sample from the control solution into a conical flask. Titrate the sodium hydroxide in the sample against standard 0.50 mol dm^{-3} hydrochloric acid solution using phenolphthalein as indicator.

Observations and calculations

Weighing

mass of weighing bottle and NaOH ....g

mass of bottle after emptying ....g

mass of NaOH ....g

mass of bottle and KHP..g

mass of bottle after emptying ....g

mass of KHP ....g

moles of KHP in 250 cm^3 of solution...moles

concentration of KHP solution ....mol dm^{-3}

Titrations
Standardising NaOH solution

run	burette reading/cm^3		titre/cm^3
	initial	final	
1			
2			
3			
		mean titre/cm^3	

amount of KHP in mean titre . . . . .mol

amount of NaOH in 20 cm^3 of solution . . . . .mol

concentration of NaOH solution . . . . .mol dm^{-3}

Back titration of NaOH after reaction with 2-ethanoyloxybenzoic acid

control	burette reading/cm^3		
run	initial	final	titre/cm^3
1			
2			
3			
		mean titre/cm^3	

aspirin	burette reading/cm^3		
run	initial	final	titre/cm^3
1			
2			
3			
		mean titre/cm^3	

amount of NaOH in control solution . . . . mol

amount of NaOH in 2-ethanoyloxybenzoic acid/NaOH solution . . . . . .mol

amount of 2-ethanoyloxybenzoic acid in each portion . . . . mol

mean mass of 2-ethanoyloxybenzoic acid for each titration . . . . .g

mass per cent of 2-ethanoyloxybenzoic acid in each tablet . . . .%

mass of 2-ethanoyloxybenzoic acid in each tablet . . . . .g

Further discussion

1   If the tablets were not 100% pure, what else could be in them?

2   How does this analysis of the tablets compare with the analysis on the back of the packet? If there are any discrepancies try to explain what may have caused them?

3   Why is it necessary to use a control solution of NaOH?

4   What sources of error are there in this experiment?

Student Guide (flow chart version)

A	**Preparing sodium hydroxide solution**
	Weighing solid sodium hydroxide
	Dissolving solid sodium hydroxide in distilled water

B	**Preparing standard potassium hydrogen phthalate solution**
	Weighing solid potassium hydrogen phthalate
	Dissolving solid potassium hydrogen phthalate in water

C	**Standardising sodium hydroxide solution**
	Titrating using potassium hydrogen phthalate solution

D	**Hydrolysis of aspirin tablets**
	Reacting with excess standardised sodium hydroxide solution

E	**Determining the purity of aspirin tablets**
	Titrating excess sodium hydroxide with hydrochloric acid solution
	Calculations

A Preparing sodium hydroxide solution

Equipment and reagents needed

- solid sodium hydroxide
- weighing bottle
- 250 cm^3 volumetric flask
- funnel
- spatula
- top pan balance

Safety

Sodium hydroxide (solid and solution): Corrosive. Causes severe burns. Risk of permanent damage to eyes.

Gloves should be worn whilst handling solid sodium hydroxide and sodium hydroxide solution.

Method

1 Weigh around 5 g of solid NaOH into a weighing bottle.

2 Transfer it to a beaker, dissolve in a small quantity of distilled water and transfer the solution to a 250 cm^3 volumetric flask.

3 Add water up to the mark, stopper and shake to mix.

mass of bottle . . . . .g

mass of bottle and NaOH . . . . .g

mass of NaOH . . . . .g

amount of NaOH . . . . .mol

approximate concentration of NaOH solution . . . . .mol dm^{-3}

B   Preparing potassium hydrogen phthalate solution

Equipment and reagents needed

- potassium hydrogen phthalate (KHP)
- weighing bottle
- 250 cm^3 volumetric flask
- funnel
- spatula
- analytical balance

Method

1 Weigh accurately (using a weighing bottle) around 12.5 g of KHP.

2 Transfer the KHP to a beaker, dissolve in a small quantity of distilled water and transfer the solution carefully into a 250 cm^3 volumetric flask.

3 Rinse the beaker several times with small quantities of distilled water and add these to the flask. Add water up to the mark with a pipette, stopper the flask and shake to mix.

mass of bottle . . . . .g

mass of bottle and KHP . . . . .g

mass of KHP . . . . .g

amount of KHP . . . . .mol

concentration of standard KHP solution .mol dm^{-3}

C   Standardising sodium hydroxide solution

Solid sodium hydroxide absorbs water from the air and so cannot easily be weighed accurately. Solutions of sodium hydroxide also slowly absorb carbon dioxide from the air. Before use, these solutions have to be standardised. A standard solution of potassium hydrogen phthalate (KHP), a monobasic acid, will be used. KHP does not absorb water from the air and can be obtained in a very pure state.

Equipment and reagents needed

- ■ sodium hydroxide solution
- ■ potassium hydrogen phthalate (KHP)
- ■ 20 cm^3 pipette
- ■ pipette filler
- ■ 3 x 100 cm^3 conical flask
- ■ 50 cm^3 burette
- ■ pipette
- ■ funnel
- ■ phenolphthalein indicator

Safety

Sodium hydroxide solution: Corrosive. Causes severe burns. Risk of permanent damage to eyes.

Phenolphthalein indicator: Highly flammable. Toxic by inhalation, in contact with skin and if swallowed. Irritating to eyes, skin and respiratory system.

Method

1   Pipette 20 cm^3 of NaOH solution into a 100 cm^3 conical flask.

2   Titrate with the KHP solution, using phenolphthalein as an indicator.

3   Repeat the titration twice.

run	burette reading/cm^3 initial	final	titre/cm^3
1			
2			
3			
		mean titre/cm^3	

D   Preparation of aspirin tablets

Aspirin is the common name for 2-ethanoyloxybenzoic acid. Boiling aspirin with sodium hydroxide solution hydrolyses the ester to yield 2-hydroxybenzoate, ethanoate anions and water.

2-ethanoyloxybenzoic acid therefore reacts with sodium hydroxide in a molar ratio of 1:2.

Equipment and reagents needed

- anti-bumping granules
- spatula
- pestle and mortar
- 50 cm^3 pipette
- pipette filler
- 2 x 250 cm^3 beaker
- hot plate
- analytical balance
- aspirin tablets

Safety

Sodium hydroxide solution: Corrosive. Causes severe burns. Risk of permanent damage to eyes.

Method

1   Weigh accurately ten aspirin tablets.

2   Grind up four aspirin tablets with a pestle and mortar.

3   Weigh accurately about 0.8 g of powdered aspirin into a beaker and add 50 cm^3 of sodium hydroxide solution from a pipette.

4   Place 50 cm^3 of sodium hydroxide into a second beaker (control solution).

5   Add a few anti-bumping granules to each beaker, then cover them both with a watch glass and boil gently for ten minutes. Allow the beakers to cool.

mass of ten aspirin tablets ....g

average mass of a single tablet ....g

mass of ground tablets treated with sodium hydroxide solution....................g

E   Determination of purity of aspirin tablets

Equipment and Reagents

- standard hydrochloric acid (0.50 mol dm^{-3})
- phenolphthalein indicator
- 10 cm^3 pipette
- pipette filler
- 50 cm^3 burette
- funnel
- 3 x 50 cm^3 conical flask
- dropper

Safety

Phenolphthalein indicator: Highly flammable. Toxic by inhalation, in contact with skin and if swallowed. Irritating to eyes, skin and respiratory system.

Method

1   Pipette 10 cm^3 of the control solution into a conical flask.

2   Titrate the excess NaOH with standard 0.50 mol dm^{-3} hydrochloric acid solution using phenolphthalein as an indicator.

3   Repeat the titration with further aliquots of the control solution.

4   Repeat steps 1–3 for the 2-ethanoyloxybenzoic acid/NaOH solution.

control	burette reading/cm^3		
run	initial	final	titre/cm^3
1			
2			
3			
		mean titre/cm^3	

aspirin	burette reading/cm^3		
run	initial	final	titre/cm^3
1			
2			
3			
		mean titre/cm^3	

mean amount of NaOH in control solution..mol

mean amount of NaOH in
2-ethanoyloxybenzoic acid/NaOH solution..mol

mean amount of aspirin in each portion ....mol

mean mass of aspirin for each titration...g

mass per cent of aspirin in each tablet...%

mass of aspirin in each tablet..g

Further discussion

1   If the tablets were not 100% pure, what else could be in them?

2   How does this analysis of the tablets compare with the analysis on the
    back of the packet? If there are any discrepancies try to explain what may
    have caused them?

3   Why is it necessary to use a control solution of NaOH?

4   What sources of error are there in this experiment?

D3

Aspirin analysis

The aim of this experiment is to calculate the purity of commercially available aspirin tablets. The procedure has three main stages – standardisation of sodium hydroxide solution; processing the aspirin tablets; and titrations to determine the purity of the tablets.

This activity is used to illustrate how a procedure may be presented in two different ways; as a traditional 'recipe' and as a flow chart, with separate pages for each part. Using a flow chart should make it easier for students to see how individual stages of the experiment relate to each other.

Pre-laboratory work

The following are a number of areas that can profitably be discussed in a pre-laboratory session. As a minimum, it is essential that students are briefed on the logic of the activity (as well as on the techniques) so that they would be able to explain this to their colleagues.

Commercial aspirin
A few different brands of aspirin, and their boxes (or photocopies of them) could be made available for students to read. Often the tablets weigh more than the mass of aspirin stated in the analysis on the back of the packet. If students are required to weigh the tablets themselves, a discussion on what the extra components of the tablets might be could be initiated. A discussion on how the drug works may also be useful.

Structure of 2-ethanoyloxybenzoic acid

The structure of 2-ethanoyloxybenzoic acid in the context of the functional groups and the reactions these might undergo will be useful. Hopefully students should notice the acid group, and realise that this will react with a base. This might prompt them to ask how a base and an ester will react and will be a good point at which to show the whole reaction scheme.

By pointing out to students that two moles of hydroxide anion react with one mole of 2-ethanoyloxybenzoic acid, it should become clear to them that this could be used as a method for the analysis of the aspirin content of tablets.

Titration

Students may suggest titrating a solution of 2-ethanoyloxybenzoic acid with sodium hydroxide directly and a discussion of why this is not suitable will be necessary. After this they might come up with the idea of reacting with excess base, and then titrating the base with an acid.

Experimental section

Drawing up a rough flow chart for the experiment as a group will ensure that students understand the reasoning behind each of the processes and the sequence that they follow. At this stage details about specific techniques are not as important as the overall scheme.

It may be necessary to demonstrate the techniques used, depending on the students' past experience of laboratory work. This point would be ideal for this, in order to ensure that students do not confuse the theory and reasoning behind the techniques with the practical details. It should be easier for them to remember how to carry out the manipulations if they know why they are carried out.

Ideas for modification and investigation

■ Standardise the hydrochloric acid with standardised sodium hydroxide rather than using commercially available standard solutions.

■ Determine the purity of paracetamol tablets.

■ Determine the relative concentrations of ingredients in tablets with more than one active ingredient.

■ Analyse tablets spectroscopically.

■ Find out what the other components of aspirin tablets are (contact the manufacturers).

Further discussion – comments

Aspirin tablets contain fillers and binders and are not one hundred per cent 2-ethanoyloxybenzoic acid. This can lead to analysis errors (in addition to the usual titration, weighing and end-point errors) if the fillers or binders react with sodium hydroxide. Use of the control sample of sodium hydroxide helps minimise systematic errors during hydrolysis.

Aspirin analysis

Equipment

- 2 x weighing bottles
- 2 x 250 cm^3 volumetric flasks
- funnel
- 10 cm^3 pipette
- 20 cm^3 pipette
- 50 cm^3 pipette
- pipette filler
- 50 cm^3 burette
- 2 x 100 cm^3 conical flasks
- 2 x 50 cm^3 conical flasks
- 250 cm^3 beaker
- glass rod
- pestle and mortar
- spatula
- plastic pipette
- analytical balance
- top pan balance
- hot plate
- anti-bumping granules
- 2 x watch glasses

Reagents

- sodium hydroxide pellets
- potassium hydrogen phthalate
- phenolphthalein indicator (1% solution in ethanol)
- aspirin tablets
- hydrochloric acid (0.50 mol dm^{-3})

4. Atomic spectra

This activity is designed to reinforce aspects of the experimental basis of atomic structure. Both the theory given and the mathematical manipulations undertaken have been simplified to improve its accessibility. The aim is to obtain a value for the Rydberg constant by measuring the frequency of the first three lines of the Balmer series for atomic hydrogen.

The manipulative work can be completed within an hour. The analysis of data and post-laboratory questions should also take little more than one hour.

This experiment is suitable for:

■ first year students

■ approximately two hours

■ individuals/pairs

Activity type

| formal | ● | experimental | ○ | divergent | ○ | investigatory | ○ |

Skills

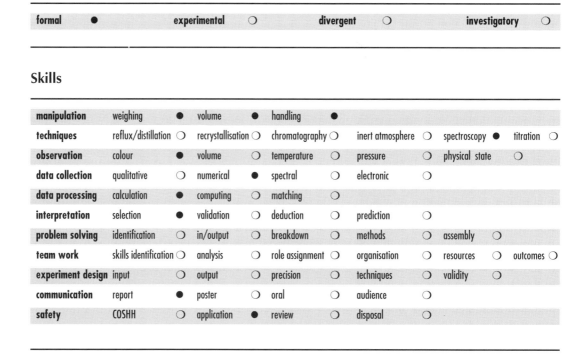

manipulation	weighing ●	volume ●	handling ●		
techniques	reflux/distillation ○	recrystallisation ○	chromatography ○	inert atmosphere ○	spectroscopy ●	titration ○
observation	colour ●	volume ○	temperature ○	pressure ○	physical state ○
data collection	qualitative ○	numerical ●	spectral ○	electronic ○	
data processing	calculation ●	computing ○	matching ○		
interpretation	selection ●	validation ○	deduction ○	prediction ○	
problem solving	identification ○	in/output ○	breakdown ○	methods ○	assembly ○
team work	skills identification ○	analysis ○	role assignment ○	organisation ○	resources ○	outcomes ○
experiment design	input ○	output ○	precision ○	techniques ○	validity ○
communication	report ●	poster ○	oral ○	audience ○	
safety	COSHH ○	application ●	review ○	disposal ○	

S4

Atomic spectra

This activity introduces atomic spectroscopy, a technique that has provided a great insight into the structure of atoms. The energy of electrons in atoms is quantised – *ie* electrons may have only certain discrete energies. Electrons can change energy by moving between energy levels and hence a photon of electromagnetic radiation is absorbed or emitted.

The energy of a photon is linked to frequency and wavelength by

$E = h\nu = hc/\lambda$

If an electron absorbs a photon of frequency ν, its energy increases by $h\nu$. Conversely, if an atom or molecule emits a photon of frequency ν, its energy decreases by $h\nu$. As there are only certain allowed energy levels for an electron in an atom or molecule, only photons of an energy corresponding to the difference between two energy levels can be absorbed or emitted. It is possible to determine the separation between energy levels in an atom or molecule by observing the frequency of radiation absorbed or emitted. Different types of energy change occur in different frequency ranges.

The separation between energy levels in the hydrogen atom can be related to n – the principle quantum number – through the Rydberg equation.

$$\bar{\nu} = R_H \left(\frac{1}{n_1^2} - \frac{1}{n_2^2} \right)$$

where

$\bar{\nu}$ is the wavenumber;

R_H is the Rydberg constant for hydrogen;

n_1 is the energy of the lower level; and

n_2 is the energy of the upper level.

The reciprocal wavelength, or wavenumber, $\bar{\nu}$, is equal to the reciprocal of the wavelength measured in cm and therefore has the units cm^{-1}.

$$\bar{\nu} = \frac{1}{\lambda} \ cm^{-1}$$

The emission spectrum of the hydrogen atom shows several series of lines that represent the energies of electronic transitions to particular levels as defined by the principle quantum number, n. For example, transitions from higher levels to the $n = 1$ level are known collectively as the Lyman series. This activity focuses on transitions to the $n = 2$ level (the Balmer series) which fall in the visible part of the spectrum.

Series	n_1	n_2	Spectral region
Lyman	1	2,3,4...	UV
Balmer	2	3,4,5...	visible
Paschen	3	4,5,6...	IR
Brackett	4	5,6,7...	IR
Pfund	5	6,7,8...	IR

Equipment

■ hydrogen discharge lamp and power supply

■ emission spectrometer

Safety

Treat all electrical equipment with respect. The discharge lamp will become hot. Do not look directly at the light.

Method

Consult a demonstrator before using the lamp or the spectrometer. Turn the hydrogen lamp on, and allow time for it to warm up (at least 20 minutes). Using the spectrometer, determine the wavelengths of the first three lines of the Balmer series, which are red, greenish-blue and violet in colour.

As n_1 is the same for all the lines observed ($n_1 = 2$), the Rydberg equation becomes

$$\overline{v} = \text{constant} - \frac{R_H}{n_2^{\,2}}$$

This function represents a straight line. By plotting $\overline{v}$ against ($1/n_2^{\,2}$), obtain values for R_H and confirm the value of n_1.

Post-laboratory work

1　Using the value for R_H calculated above, predict where the fourth line of the Balmer series would be expected.

2　How does your value for R_H compare with the literature value of 1.09678×10^5 cm^{-1}? If your answer is significantly different, is there any reason for the deviation? To what level of precision is your answer?

3　Using the experimental value of R_H, calculate the wavenumbers of the first three lines in the Lyman series for atomic hydrogen.

Atomic spectra

This activity introduces students to atomic spectroscopy. They are asked to find frequency values for the first three lines of the Balmer series. By plotting a graph of wavenumber against $1/n_2^{\,2}$, students then obtain an experimental value for the Rydberg constant (R_H) and $R_H/n_1^{\,2}$. In the post-laboratory session students use these values to predict the wavenumber of the fourth line in the Balmer series, and compare the values they have obtained with literature values.

Pre-laboratory work

Students must read through the entire experimental schedule before the pre-laboratory session. After sorting out any problems, the following could be discussed:

1   What is the energy per photon of a 60 W yellow light bulb (assume $\lambda = 550$ nm)? How many photons are emitted per second? What assumptions have been made?

Each photon has energy $h\nu = \dfrac{hc}{\lambda}$

The total energy emitted by the 60 W (Js^{-1}) yellow light bulb per second is 60 J

The total number of photons needed to produce energy E is $\dfrac{E\lambda}{hc}$

number of photons emitted per second $= \dfrac{550 \times 10^{-9} \times 60}{6.626 \times 10^{-34} \times 3 \times 10^{8}}$

$= 1.7 \times 10^{20}$

energy per photon $= \dfrac{60}{1.7 \times 10^{20}} = 3.6 \times 10^{-19}$ J

2   Construct a simple energy level diagram showing the first three transitions in each of the Lyman, Balmer and Paschen series.

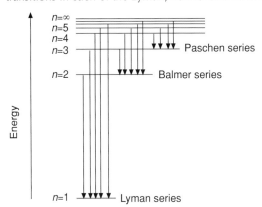

3   Does an orbital exist if there is not an electron in it?

An atomic orbital is a one-electron wavefunction for an electron in an atom, and is defined by three quantum numbers. When such a wavefunction describes one of these electrons it is said to 'occupy' that orbital.

4   Does the Schrödinger wave equation allow the calculation of exact energy levels for hydrogenic atoms? What factors prevent the exact calculation of energy levels for non-hydrogenic atoms?

Note: hydrogenic atoms are one-electron atoms or ions of general atomic number **Z**, for example H, He$^+$, Li^{2+} or U^{91+}.

The Schrödinger equation does allow calculation of energy levels for one-electron species. However non-hydrogenic species have many electrons which interact with one another. This means that approximations have to be made and no analytical expressions for the orbitals and energies can be given.

It may be effective to get the students to answer questions as a group. As some of the theory has not been covered in the experimental introduction, further guidance and suggestions for relevant textbooks may be needed. Some of the questions are deliberately more discursive.

Error analysis

By pooling data for the class, the standard deviation, s, can be calculated using the formula:

$$s = \sqrt{\frac{1}{n-1} \sum_{i=1}^{n} (x_i - \bar{x})^2}$$

where

x_i = individual result

$\bar{x}$ = mean of sample population

n = number of samples taken

Post-laboratory work

Literature values for the constants calculated by students are as follows: R_H = 109 678 cm^{-1} and first four lines of the Balmer series 15 228 cm^{-1}, 20 571 cm^{-1}, 23 638 cm^{-1} and 24 380 cm^{-1}.

Literature values for the first three bands of the Lyman series are as follows: 82 258 cm^{-1}, 97 491 cm^{-1} and 102 824 cm^{-1}.

Further investigation

Students can further their investigation by measuring the visible line spectrum of sodium, potassium or helium, and use the data to estimate their ionisation energies. The spectrometer can also be used to measure the temperature of a sodium flame.

Atomic spectra

Equipment

- hydrogen discharge lamp and power supply
- emission spectrometer

5. What's in a solution?

This experiment is designed to improve observation, deduction and communication skills. There are three different activities – the first requires students to record their observations, and the second and third require them to rationalise their observations and hence identify each solution.

This experiment is suitable for a wide range of ability levels. For the first two parts very little chemical knowledge is required. For the last part, students must have some knowledge of the solubilities and colours of salts.

This experiment is suitable for:

■ a wide range of ability levels

■ 1 hour

■ individuals/pairs

Activity type

| formal | ○ | experimental | ● | divergent | ○ | investigatory | ○ |

Skills

manipulation	weighing	○	volume	○	handling	●						
techniques	reflux/distillation	○	recrystallisation	○	chromatography	○	inert atmosphere	○	spectroscopy ●		titration	○
observation	colour	●	volume	○	temperature	●	pressure	○	physical state ●			
data collection	qualitative	●	numerical	●	spectral	○	electronic	○				
data processing	calculation	●	computing	○	matching	○						
interpretation	selection	○	validation	○	deduction	●	prediction	○				
problem solving	identification	●	in/output	○	breakdown	○	methods	○	assembly	●		
team work	skills identification	○	analysis	○	role assignment	○	organisation	○	resources	○	outcomes	○
experiment design	input	○	output	○	precision	○	techniques	○	validity	○		
communication	report	●	poster	○	oral	○	audience	○				
safety	COSHH	○	application	●	review	○	disposal	○				

S5

What's in a solution?

In this experiment the skills of observation and deduction are developed. Firstly the solutions are mixed in a beaker as indicated, and any observations recorded. Secondly the observations are used to identify each of the solutions.

Equipment

- 250 cm^3 beaker
- 10 cm^3 measuring cylinder
- stirring rod
- dropping pipettes
- acetate sheets

Reagents

Part 1
- cobalt(II) nitrate solution (0.10 mol dm^{-3})
- zinc(II) nitrate solution (0.10 mol dm^{-3})
- copper(II) sulfate solution (0.50 mol dm^{-3})
- potassium thiocyanate (0.50 mol dm^{-3})
- pyridine

Part 2
Solutions A–D are:
- hydrochloric acid (2.0 mol dm^{-3})
- sodium hydroxide (2.0 mol dm^{-3})
- phenolphthalein solution
- water

Part 3
Solutions A–G are:
- copper(II) sulfate solution (0.10 mol dm^{-3})
- nickel(II) sulfate solution (0.10 mol dm^{-3})
- barium chloride solution (0.10 mol dm^{-3})
- potassium chloride solution (0.10 mol dm^{-3})
- potassium iodide solution (0.10 mol dm^{-3})
- copper(II) chloride solution (0.10 mol dm^{-3})
- copper(II) chloride solution (0.50 mol dm^{-3})

Safety

Copper(II) sulfate solution: Harmful if swallowed. Irritating to skin and eyes.

Potassium thiocyanate: Harmful by ingestion, inhalation and skin contact. Irritating to eyes and skin.

Pyridine: Highly flammable. Harmful by ingestion, inhalation and skin contact.

Hydrochloric acid: Irritating to eyes, skin and respiratory system. Causes burns.

Sodium hydroxide: Destructive to eyes and skin. Causes severe burns.

Phenolphthalein solution: Highly flammable. Toxic by inhalation, in contact with skin and if swallowed. Irritating to eyes, skin and respiratory system.

Nickel(II) sulfate solution: Prolonged contact with skin can cause dermatitis.

Barium chloride: Harmful by inhalation and if swallowed. Will decompose violently on heating.

Potassium chloride: Harmful. Explosive if mixed with combustible material.

Potassium iodide: May cause sensitisation by inhalation and skin contact. Possible risk of harm to unborn child. Irritating to skin, eyes and respiratory system.

Copper(II) chloride: Harmful by ingestion.

In the case of contact of hydrochloric acid, sodium hydroxide or barium chloride with skin or eyes, wash with copious amounts of water. Gloves should be worn when handling strong acids and alkali.

Observations

The way in which observations are recorded is important – using the record you produce someone else should be able to repeat the experiment and know if they have achieved the same results. For each section a record of observations and an approximate timescale should be noted, especially if changes do not take place immediately. If there is any doubt as to whether something is important enough to record it should be recorded, as it can always be edited out of a final report.

Part 1

Record all observations, without trying to rationalise them.

1 Place 200 cm^3 of water into a beaker, and add 10 cm^3 of cobalt nitrate solution, 10 cm^3 of pyridine, and finally 10 cm^3 of potassium thiocyanate solution. Leave the mixture to stand for around 10 minutes.

2 Place 200 cm^3 of water into a beaker, and add 10 cm^3 of zinc nitrate solution, 10 cm^3 of pyridine and 10 cm^3 of potassium thiocyanate solution.

3 Place 200 cm^3 of water into a beaker, and add 9.5 cm^3 of zinc nitrate solution, 0.5 cm^3 cobalt nitrate solution, 10 cm^3 of pyridine and 10 cm^3 of potassium thiocyanate solution.

4 Place 200 cm^3 of water into a beaker, and add one drop of copper sulfate solution, 10 cm^3 of pyridine and 10 cm^3 of potassium thiocyanate solution. Changes in this mixture will take place very slowly.

5 Place 200 cm^3 of water into a beaker, and add 10 cm^3 of zinc nitrate solution, one drop of copper sulfate solution, 10 cm^3 of pyridine and 10 cm^3 of potassium thiocyanate solution.

Parts 2 and 3

Solutions labelled **A–D** for Part 2 and **A–G** for Part 3 are provided. Without using any other reagents identify each solution. Small volumes of solutions can be easily mixed and observed by dropping them onto an acetate sheet, which has been divided into small squares. If necessary, place the acetate sheet onto a piece of white paper to aid observation. Present the results for each part in a separate table, detailing any observations and identifying each solution. Give a brief explanation for your observations.

What's in a solution?

The aim of this experiment is to develop observation skills, and to identify a number of different solutions. For the first part no rationalisation of observations is expected. In the second and third parts the solutions need to be mixed with each other, and the observations rationalised to make the identification of each solution possible. Neither part of the experiment requires much chemical knowledge, as it is the thought processes which are most important here. However, the second part does require some knowledge of the colours and solubilities of salts.

This experiment develops students' skills in problem solving, experimental design, logical thinking, organisation of work and observation.

Part 1

1   A pink precipitate appears quite slowly and has the formula $[Co(py)_2(NCS)_2]$. Precipitation is usually complete after ten minutes. The precipitate can be converted to the grey-pink polymeric form of $[Co(py)_2(NCS)_2]$ by heating in an oven at 100 °C for approximately one hour.

2   A white precipitate forms almost immediately, which has formula $[Zn(py)_2(NCS)_2]$.

3   A white precipitate forms almost immediately, but as crystals grow they become blue in colour.

4   A small amount of green-blue precipitate forms very slowly.

5   A white precipitate forms almost immediately, but as crystals grow they turn violet in colour.

Part 2

When the base and phenolphthalein are mixed together, there will be a colour change from colourless to pink or red. If water is then added to this mixture, the colour will not change, but will become more diluted. If acid is added, the colour will fade. These observations identify all of the solutions, apart from differentiating the base from the phenolphthalein. If the acid and base are mixed together in equal proportions, then adding phenolphthalein to this mixture should result in no colour change, regardless of how much phenolphthalein is added. On the other hand, if base were added to a mixture of acid and phenolphthalein, then the colour would change to pink as more base was added.

Part 3

The two blue solutions are either copper(II) sulfate or dilute copper(II) chloride. If these are mixed with the colourless solutions, the observations should be:

	$CuSO_4$	$CuCl_2$
$BaCl_2$	white precipitate	goes green
KCl	goes green	goes green
KI	white precipitate/ violet precipitate	white precipitate/ violet precipitate

Add the barium chloride to the green solutions. The observations should be:

	$NiSO_4$	$CuCl_2$
$BaCl_2$	white precipitate	stays green

The green colour in the concentrated copper(II) chloride solution is due to the $[CuCl_4]^{2-}$ ion which is formed in concentrated solution. It is also formed when chlorine is added to the solution.

T5

What's in a solution?

Equipment

- 250 cm^3 beaker
- 10 cm^3 measuring cylinder
- stirring rod
- dropping pipettes
- acetate sheets

Reagents

Part 1
- cobalt(II) nitrate solution (0.10 mol dm^{-3})
- zinc(II) nitrate solution (0.10 mol dm^{-3})
- copper(II) sulfate solution (0.50 mol dm^{-3})
- potassium thiocyanate (0.50 mol dm^{-3})
- pyridine

Part 2
The following solutions should be labelled A–D:
- hydrochloric acid (2.0 mol dm^{-3})
- sodium hydroxide (2.0 mol dm^{-3})
- phenolphthalein solution
- distilled water

Part 3
The following solutions should be labelled A-G:
- copper(II) sulfate solution (0.10 mol dm^{-3})
- nickel(II) sulfate solution (0.10 mol dm^{-3})
- barium chloride solution (0.10 mol dm^{-3})
- potassium chloride solution (0.10 mol dm^{-3})
- potassium iodide solution (0.10 mol dm^{-3})
- copper(II) chloride solution (0.10 mol dm^{-3})
- copper(II) chloride solution (0.50 mol dm^{-3})

6. Organic identification

This is another 'puzzle' type of organic chemistry experiment. An unknown glycol is treated with sulfuric acid. The boiling point of the product is recorded, and infrared and proton NMR spectra of both the product and the starting material are collected. Using all of the information collected it is then possible to identify the unknown glycol.

The manipulative part of this experiment is relatively straightforward, and first year students with only limited experience in the laboratory should be able to tackle it successfully. The work should take approximately two to three hours.

This activity is adapted from a procedure by B J Wojciechowski and Todd S Deal, *J. Chem. Educ.*, 1996, **73**, 85.

This experiment is suitable for:

■ first year students

■ two to three hours

■ individuals/pairs

Activity type

formal	○	experimental	●	divergent	○	investigatory	○

Skills

manipulation	weighing	●	volume	○	handling	●						
techniques	reflux/distillation	●	recrystallisation	○	chromatography	○	inert atmosphere	○	spectroscopy	●	titration	○
observation	colour	○	volume	○	temperature	●	pressure	○	physical state	○		
data collection	qualitative	○	numerical	●	spectral	●	electronic	○				
data processing	calculation	○	computing	○	matching	●						
interpretation	selection	●	validation	○	deduction	●	prediction	○				
problem solving	identification	●	in/output	○	breakdown	○	methods	○	assembly	●		
team work	skills identification	○	analysis	○	role assignment	○	organisation	○	resources	○	outcomes	○
experiment design	input	○	output	○	precision	○	techniques	○	validity	○		
communication	report	●	poster	○	oral	○	audience	○				
safety	COSHH	○	application	●	review	○	disposal	○				

S6

Organic identification

The aim of this experiment is to carry out a reaction on an unknown sample, and then use spectroscopic data to determine the structures of both the starting material and product, and the nature of the reaction.

Equipment

- 50 cm^3 measuring cylinder
- 100 cm^3 round-bottomed flask
- condenser
- heating mantle
- 2 x 50 cm^3 beaker
- dropping pipettes
- spatula
- fractional distillation column
- thermometer

Reagents

- unknown glycol
- concentrated sulfuric acid
- anhydrous magnesium sulfate

Safety

Unknown glycol: Irritating to eyes, skin and respiratory system.

Concentrated sulfuric acid: Burns skin and eyes. Causes severe damage if taken by mouth. Toxic by inhalation of fumes or mist.

Anhydrous magnesium sulfate: Harmful by inhalation, in contact with skin and if swallowed.

Wear gloves when handling strong acid, and in case of contact with eyes or skin wash with copious amounts of water.

Method

A sample of a glycol of molecular formula $C_6H_{14}O_2$ is provided. First record its infrared and ^{1}H NMR spectra.

Place 40 cm^3 of water into a 100 cm^3 round-bottomed flask. Add 10 cm^3 of concentrated sulfuric acid, and 10.0 g of the solid glycol. Reflux the solution for 15 minutes.

Allow the solution to cool, and then distil it until the upper level of the distillate no longer increases. Separate the upper layer, dry over anhydrous magnesium sulfate, and purify by fractional distillation. Record the boiling point of the product and obtain an infrared spectrum and a ^{1}H NMR spectrum of the product.

Report

Record details of the experimental procedure, including quantities of reagents used and the product yield, in the report.

Use the spectra to identify the main characteristics of both the starting material and product.

Use the boiling point and spectroscopic information to identify both the product and starting material.

Suggest a possible mechanism for the reaction carried out.

D6

Organic identification

The aim of this experiment is to carry out a reaction on an unknown sample, and then use spectroscopic data to determine the structures of both the starting material and product, and the nature of the reaction.

This experiment is based on the pinacol rearrangement. Students are given a sample of 2,3-dimethyl-2,3-butanediol (pinacol), along with a procedure for the reaction, but are only told that their starting material is a simple glycol of molecular formula $C_6H_{14}O_2$. They are expected to obtain infrared and ^{1}H NMR spectra of both the starting material and the product (see attached figures), and use these, together with the boiling point of the product to determine their structures. They are then expected to propose a plausible mechanism for the reaction.

The boiling point of 3,3-dimethyl-2-butanone is 106 °C.

The mechanism of the reaction, shown below, involves a simple 1,2-shift.

The experimental procedure is quite simple, and is an ideal way of introducing fractional distillation to students. If they have not previously recorded their own spectra, or if time is short, it is possible to supply them with spectra of the starting material and the product. It is recommended, however, that students record at least an IR spectrum of their product, otherwise there is little point in them carrying out the reaction. This also serves as a good way of letting them see for themselves how pure their product is.

T6

Organic identification

Equipment

- 50 cm^3 measuring cylinder
- 100 cm^3 round-bottomed flask
- condenser
- heating mantle
- 2 x 50 cm^3 beaker
- dropping pipettes
- spatula
- fractional distillation column
- thermometer

Reagents

- 2,3-dimethyl-2,3-butanediol (labelled as the unknown glycol)
- concentrated sulfuric acid
- anhydrous magnesium sulfate

7. Properties of a zeolite

In this activity, zeolite A is extracted from washing powder. Its properties are studied in an open investigation.

This experiment is suitable for:

■ first or second year students

■ approximately three hours

■ groups or individuals

Activity type

formal	○	experimental	○	divergent	●	investigatory	○

Skills

manipulation	weighing	●	volume	○	handling	●						
techniques	reflux/distillation	○	recrystallisation	○	chromatography	○	inert atmosphere	○	spectroscopy	○	titration	○
observation	colour	●	volume	○	temperature	○	pressure	○	physical state	●		
data collection	qualitative	●	numerical	●	spectral	○	electronic	○				
data processing	calculation	○	computing	○	matching	●						
interpretation	selection	●	validation	○	deduction	●	prediction	○				
problem solving	identification	○	in/output	○	breakdown	○	methods	●	assembly	○		
team work	skills identification	○	analysis	○	role assignment	○	organisation	○	resources	○	outcomes	○
experiment design	input	○	output	○	precision	○	techniques	●	validity	○		
communication	report	●	poster	○	oral	○	audience	○				
safety	COSHH	○	application	●	review	○	disposal	○				

S7

Properties of a zeolite

In this activity, zeolite A is extracted from washing powder and some of its properties are investigated.

Equipment

- Buchner flask and funnel
- filter paper
- 2 x large test-tubes
- 2 x stoppers
- oven or furnace at 400 °C
- analytical balance
- burette
- conical flask
- pipette
- glass wool
- Pyrex glass tubing
- tube furnace or resistance heated electrical tape

Reagents

- washing powder
- ammonium chloride solution (10 g NH_4Cl in 40 cm^3 H_2O)
- liquid dish soap
- calcium(II) chloride
- ethylenediaminetetraacetic acid (EDTA)
- nitrogen gas supply
- mineral oil
- 2,2-dimethylpropan-1-ol (neopentyl alcohol)

Safety

Ammonium chloride solution: Harmful if swallowed. Irritating to eyes.

Calcium chloride: Irritating to eyes.

EDTA: Irritating to eyes, skin and respiratory system.

2,2-dimethylpropan-1-ol: Flammable.

Zeolite A (Na-A) is used in many laundry detergents as a water softener. It softens water by cation exchange of its sodium ions for calcium and magnesium ions. However, the applications of zeolites extend far beyond water softening as they can be used as desiccants, selective adsorbents and as highly selective catalysts.

Na-A is the only insoluble component of detergent, and hence can be isolated by filtering a solution of the detergent followed by washing with water. Having isolated Na-A, determine its weight percentage in the detergent. As it is hygroscopic, the powder should be heated to 400 °C before weighing.

Choose two of the following areas to investigate.

Water softening properties of Na-A
Design an experiment to show the water-softening property of Na-A.

Desiccant properties of Na-A
Determine how much water dry Na-A will absorb. Calculate this as percentage by mass.

Ion exchange properties of Na-A
Detergent zeolite has an approximate empirical formula of $NaAlSiO_4$. If Na-A is placed in a solution rich in Ca^{2+} ions then ion exchange yields a zeolite with the formula $Ca_{0.5}AlSiO_4$. Design an experiment to determine the degree of ion exchange.

Use as a dehydration catalyst
Na-A catalyses the dehydration of an alcohol to form an alkene. To make the zeolite catalytically active it is converted to its acid form (H-A). The Na^+ ions in Na-A are first exchanged for NH_4^+ ions by stirring a few grams of zeolite in an ammonium chloride solution for one hour. The supernatant solution should be discarded, and the zeolite filtered and washed. The NH_4^+ zeolite should then be heated in air overnight at 400 °C to drive off ammonia and produce the protonated form, H-A.

Place a few grams of H-A in the middle section of a length of Pyrex glass tubing, and hold it in place with a loose plug of glass wool at each end. Connect one end of the tube to the outlet of a bubbler containing 2,2-dimethylpropan-1-ol, whose inlet is connected to a nitrogen cylinder. The other end of the Pyrex tubing should be connected to a second bubbler containing mineral oil. Wrap the glass with resistance heated electrical tape, or heat it in a tube furnace to around 300 °C. Pass a slow flow of nitrogen through the apparatus. The alkene is collected as a solution in the mineral oil bubbler.

Take an infrared spectrum of the oil solution to verify that it does in fact contain an alkene. What is the identity of this alkene? Write a balanced equation for the reaction carried out.

Properties of a zeolite

In this activity, zeolite A is extracted from washing powder and some if its properties are investigated.

The isolation of the zeolite is a simple procedure, and students should have little difficulty with it. As the zeolite crystals are very small (~1 μm) a fine filter paper should be used.

Water softening properties of Na-A

This can be demonstrated by shaking a large stoppered test-tube half-filled with tap-water, around 0.25 g of Na-A and a small drop of liquid dish soap. The height of the suds layer formed in the test-tube will be greater than in a similar tube containing soapy water but no Na-A.

Desiccant properties of Na-A

This can be quantified by first weighing a few grams of zeolite after drying overnight at 400 °C, then placing the zeolite in a humid environment and reweighing until the mass has equilibrated. Na-A can absorb approximately 20% of its weight in water.

Ion exchange properties of Na-A

The Ca^{2+} solution should be prepared by dissolving 15 g $CaCl_2.H_2O$ in 200 cm^3 of distilled water. The solution should be standardised by titration of Ca^{2+} with EDTA. Students should experiment to find how much Na-A to add to the solution, and how long to leave it. They should titrate an aliquot of the solution both before and after addition of Na-A.

Zeolite H–A as a dehydration catalyst

Students have been given more detailed instructions about this part of the experiment as it is more complex. They may need help assembling all of the equipment properly.

The salient feature of the infrared spectrum should be the olefinic C–H out-of-plane wagging vibration at 801 cm^{-1}. There is a C=C stretch at 1660 cm^{-1} but it is very weak. This reaction should produce 2-methyl-2-butene and water.

T7

Properties of a zeolite

Equipment

- Buchner flask and funnel
- filter paper
- 2 x large test-tubes
- 2 x stoppers
- oven or furnace at 400 °C
- analytical balance
- burette
- conical flask
- pipette
- tube furnace or resistance heated electrical tape
- glass wool
- Pyrex glass tubing

Reagents

- washing powder
- ammonium chloride solution (10 g NH_4Cl in 40 cm^3 H_2O)
- liquid dish soap
- calcium(II) chloride
- ethylenediaminetetraacetic acid (EDTA)
- nitrogen gas supply
- mineral oil
- 2,2-dimethylpropan-1-ol

8. Nicotine and smoking

In this activity, students extract nicotine from tobacco, and investigate the concentration of their extracts using gas chromatography. They then relate this to the amount of nicotine in the original sample.

This experiment is suitable for:

■ first or second year students

■ approximately three hours

■ individuals/pairs

Activity type

formal	○	experimental	○	divergent	○	investigatory	●

Skills

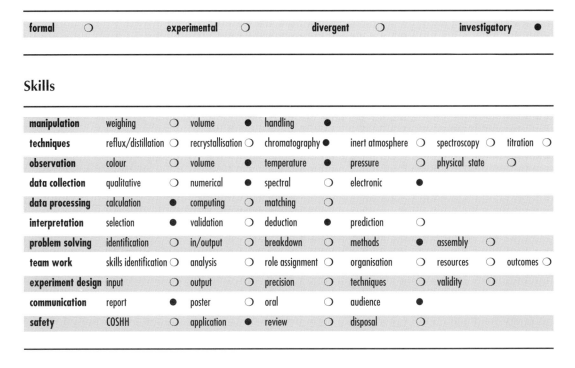

manipulation	weighing	○	volume	●	handling	●						
techniques	reflux/distillation	○	recrystallisation	○	chromatography	●	inert atmosphere	○	spectroscopy	○	titration	○
observation	colour	○	volume	●	temperature	●	pressure	○	physical state	○		
data collection	qualitative	○	numerical	●	spectral	○	electronic	●				
data processing	calculation	●	computing	○	matching	○						
interpretation	selection	●	validation	○	deduction	●	prediction	○				
problem solving	identification	○	in/output	○	breakdown	○	methods	●	assembly	○		
team work	skills identification	○	analysis	○	role assignment	○	organisation	○	resources	○	outcomes	○
experiment design	input	○	output	○	precision	○	techniques	○	validity	○		
communication	report	●	poster	○	oral	○	audience	●				
safety	COSHH	○	application	●	review	○	disposal	○				

S8

Nicotine and smoking

In this activity the amount of nicotine in a variety of different types of tobacco is measured.

Equipment

- 2 x 4 cm^3 sample vials, with caps and septa
- sonicating bath
- top pan balance
- gas chromatograph
- 1 µl syringe

Reagents

- potassium hydroxide solution in methanol (0.05 mol dm^{-3})
- nicotine standard solution
- tobacco samples

Safety

Potassium hydroxide solution in methanol: Highly flammable. Corrosive.

Methanol: Toxic by ingestion, inhalation and skin contact. Can cause delayed damage to eyes if ingested.

Nicotine solution: Highly toxic.

All solutions containing nicotine should be handled with gloves.

Extracting nicotine from tobacco

Using a top pan balance weigh out precisely a 30–40 mg quantity of each tobacco into separate 4 cm^3 sample vials. Add 2.0 cm^3 of methanolic potassium hydroxide solution to each sample. Cap the vials. Place in the sonicating bath for 30 minutes.

Calibrating the gas chromatograph

A nicotine standard solution will be provided with which to calibrate the machine. Other students will analyse standards of different concentrations, so that as a team a calibration graph for the instrument can be constructed. If more than one instrument is available, each instrument will need to be calibrated separately.

Inject a 0.5 μl sample of the nicotine standard provided into the gas chromatograph. The trace will take approximately 10 minutes.

Plot a graph of peak area against nicotine concentration. This is your calibration graph.

Analysing the tobacco extracts

Inject a 0.5 μl sample of one of the extracts into the gas chromatograph. When the trace is completed, clean the syringe, and inject a 0.5 μl sample of the other extract. Use the calibration chart to calculate the nicotine content of each extract. The concentrations determined are expressed in mg cm^{-3}, but as each sample was extracted with 2.0 cm^3 of solvent, multiply these values by 2.0 to calculate the mass of nicotine in the extracts. Express the results as milligrams of nicotine per milligram of tobacco.

Post-laboratory questions

1   An average cigarette contains about 700 mg of tobacco. Calculate the nicotine content of an average cigarette of the tobacco analysed.

2   How does the value determined experimentally compare with the value printed on the cigarette packet? What are the reasons for any differences found?

RS•C

Nicotine and smoking

In this experiment the amount of nicotine in a variety of tobaccos is measured.

Allocation of tobacco and nicotine standards

At least three different standard nicotine solutions should be made up. Each student should be assigned one standard solution to use for calibrating the instrument.

Collect a variety of cigarette, cigar and loose tobaccos. Students should be given two samples to analyse, one of which should be cigarette tobacco.

Gas chromatography

If students are not familiar with gas chromatographs, they must be shown how to use the instrument and clean the syringes. Ensure that once a student has chosen an instrument to work with, they continue to use the same instrument for the rest of the experiment. The following machine settings should be used:

Gas settings:
- N_2 carrier   35 psi ~ 30 cm^3 min^{-1}
- H_2 10 psi ~ 30 cm^3 min^{-1}
- Air   25 psi ~ 300 cm^3 min^{-1}

Temperature:
- Column   180 °C
- Detector   250 °C
- Injector   200 °C

Post-laboratory questions

1 The nicotine content of cigarette tobacco seems to lie in the range 0.010–0.030 mg nicotine/mg tobacco. A cigarette of *ca* 700 mg may typically have a nicotine content of 10 mg.

2 The value that students have calculated should be in excess of the value of 1 mg quoted on most packets. The discrepancy lies in the way the value quoted on packets is determined. This value comes from a smoking machine, in which the smoke is passed through a filter on which the nicotine is trapped. It is the value of the 'trapped' nicotine that is determined and quoted on the cigarette packet.

Nicotine and smoking

Reagents

- potassium hydroxide solution in methanol
- a range of nicotine standard solutions*
- tobacco samples

Equipment

- 2 x 4 cm^3 sample vials, with caps and septa
- sonicating bath
- top pan balance
- 1 μl syringe
- gas chromatograph

*The range of standard nicotine solutions provided to students will be used to calibrate the gas chromatograph and should be chosen to be comparable with the equipment used.

9. Infrared spectroscopy

The aim of this activity is to develop students' understanding of the basis of infrared spectroscopy, and to introduce some of the techniques involved in running spectra. A pre-laboratory session is undertaken before students are shown how to operate the spectrometer. The student guide details an analysis of the spectra which is undertaken after they have been recorded.

This experiment is suitable for:

■ first year students

■ approximately three hours

■ individuals/pairs

Activity type

| formal | ● | experimental | ○ | divergent | ○ | investigatory | ○ |

Skills

manipulation	weighing	○	volume	○	handling	●				
techniques	reflux/distillation	○	recrystallisation	○	chromatography	○	inert atmosphere ○	spectroscopy ●	titration ○
observation	colour	○	volume	○	temperature	○	pressure ○	physical state ○
data collection	qualitative	○	numerical	●	spectral	●	electronic ○
data processing	calculation	●	computing	○	matching	○
interpretation	selection	○	validation	●	deduction	●	prediction ●
problem solving	identification	○	in/output	○	breakdown	○	methods ●	assembly ○
team work	skills identification	○	analysis	○	role assignment	○	organisation ○	resources ○	outcomes ○
experiment design	input	○	output	○	precision	○	techniques ○	validity ○
communication	report	○	poster	○	oral	●	audience ●
safety	COSHH	○	application	●	review	○	disposal ○

RS•C

57

Infrared spectroscopy

This activity introduces some important concepts of infrared spectroscopy. Initially a number of questions about the theory of infrared spectroscopy are posed, and then spectra are recorded and analysed.

Pre-laboratory work

Infrared (IR) spectroscopy is used to investigate the vibrational properties of molecules. It can lead to the unambiguous identification of simple molecules, and can reveal important structural and bonding information about more complex molecules. IR spectroscopy has applications in inorganic, organic and physical chemistry, aside from obvious analytical applications.

Electromagnetic radiation in the infrared region interacts with molecular vibrations of the same frequency. This results in a molecule vibrating at the same frequency, but with an increased amplitude. A detector is used to find the frequency of light which is absorbed by the sample. In order for a molecular vibration to be infrared active, the electric dipole moment associated with it must change.

The frequency of radiation absorbed can be related to the strength of the bond, and the reduced mass of the bonded atoms by the equation:

$$\nu = \frac{1}{2\pi} \sqrt{\frac{k}{\mu}}$$

(1)

where

ν is the frequency of radiation absorbed;

k is the strength of the bond; and

μ is the reduced mass of the bonded atoms.

The reduced mass of two atoms, A and B, is equal to:

$$\mu = \frac{m_A m_B}{m_A + m_B}$$

(2)

where m_A and m_B are the masses of A and B respectively.

Using equation (2), calculate the reduced mass of an HCl molecule. Note that the reduced mass must be in kg.

Reagents

- trichloromethane
- deuterated trichloromethane
- propan-1-ol
- *n*-heptane

Equipment

- infrared spectrophotometer and cells

Safety

Trichloromethane: Harmful if swallowed. Irritating to skin. Danger of cumulative effects. Carcinogen.

propan-1-ol: Highly flammable. Harmful by ingestion and inhalation. Irritating to eyes and skin.

***n*-heptane:** Highly flammable. Harmful by ingestion, inhalation and skin contact. Degreases skin. Irritating to eyes, skin and respiratory system.

Infrared spectrum of $CHCl_3$ and $CDCl_3$

1   Record the infrared spectrum of $CHCl_3$ in the range 600 cm^{-1} to 3500 cm^{-1}.

2   Identify the C–H stretch, and use the value of its frequency to calculate k. It can be assumed that the value of k does not change substantially between a C–H and a C–D bond, if nothing else about the molecule is changed.

3   Use the value of k obtained to predict the frequency of the C–D stretch in $CDCl_3$.

4   Now record the infrared spectrum of $CDCl_3$ in the same range and identify the C–D stretch.

5   Calculate the percentage deviation between the predicted and observed frequencies.

The effect of hydrogen bonding on vibrational frequencies

1   Record the infrared spectra of 0.1 mol dm^{-3} and 0.6 mol dm^{-3} solutions of propan-1-ol in *n*-heptane in the range 3000–4000 cm^{-1}.

2   There should be two O–H stretching bands, at 3640 cm^{-1} and 3350 cm^{-1}. Find these peaks on both spectra and label them.

3   Why are there two O–H stretches for propan-1-ol when there is only one O–H bond in the molecule?

4   Why does the ratio of these peaks change with the propan-1-ol concentration?

References

W B Heuer and E Koubek, *J. Chem. Educ.*, 1997, **74**, 313–315.

J S Anderson, D M Hayes and T C Weiner, *J. Chem. Educ.*, 1995, **72**, 653–655.

D9

Infrared spectroscopy

The aim of this activity is to develop students' understanding of infrared spectroscopy, and to teach them how to run and analyse spectra.

Pre-laboratory work

$$m_H = \frac{1.00794}{6.022 \times 10^{23}} = 1.67376 \times 10^{-27} \text{ kg}$$

$$m_{Cl} = \frac{35.4527}{6.022 \times 10^{23}} = 5.8872 \times 10^{-26} \text{ kg}$$

$$\mu = \frac{1.67376 \times 10^{-27} \times 5.8872 \times 10^{-26}}{1.67376 \times 10^{-27} + 5.8872 \times 10^{-26}} \text{ kg}$$

$$= 1.6275 \times 10^{-27} \text{ kg}$$

After students have completed the question in the pre-laboratory section, a demonstration of how to run spectra (using $CHCl_3$) should be given. Some guidance on how to identify the major bands of the spectra might also be given.

Students should then work through the experiment in pairs.

Infrared spectra of $CHCl_3$ and $CDCl_3$

Students may need help identifying the C–H stretching band. The C–H band should be located at 3020 cm^{-1} and the C–D band at 2254 cm^{-1}. Predicted C–D stretching frequencies are typically 3–5% lower than those observed.

The effect of hydrogen bonding on vibrational frequencies

There are two O–H stretching bands in the spectrum of propan-1-ol because of hydrogen bonding. The ratio of peaks changes because there is a shift from free O–H stretching at 3640 cm^{-1} to hydrogen bonded O–H stretching at 3350 cm^{-1} as the concentration of propan-1-ol increases.

T9

Infrared spectroscopy

Equipment

■ infrared spectrophotometer and cells

Reagents

■ trichloromethane

■ deuterated trichloromethane

■ propan-1-ol

■ *n*-heptane

LIVERPOOL JOHN MOORES UNIVERSITY
LEARNING SERVICES

10. Synthesis and analysis of an aluminium oxalato compound

Metal complexes are usually associated with the transition elements but many such compounds of the main group elements also exist. This activity involves the synthesis of an aluminium oxalato complex from a solution of potassium aluminate. The subsequent analysis is carried out titrimetrically with the oxidation of the oxalato group to carbon dioxide using manganate(VII) solution.

This experiment is suitable for:

■ first or second year students

■ approximately three hours

■ group or individual work

Activity type

formal	○	experimental	○	divergent	○	investigatory	●

Skills

manipulation	weighing	●	volume	●	handling	●				
techniques	reflux/distillation	○	recrystallisation	●	chromatography	○	inert atmosphere	○	spectroscopy ○ titration ●
observation	colour	○	volume	●	temperature	○	pressure	○	physical state ●
data collection	qualitative	●	numerical	●	spectral	○	electronic	○	
data processing	calculation	●	computing	○	matching	●			
interpretation	selection	○	validation	●	deduction	●	prediction	○	
problem solving	identification	●	in/output	●	breakdown	○	methods	●	assembly ○
team work	skills identification	○	analysis	○	role assignment	○	organisation	○	resources ○ outcomes ○
experiment design	input	○	output	○	precision	○	techniques	●	validity ○
communication	report	●	poster	○	oral	○	audience	○	
safety	COSHH	○	application	●	review	○	disposal	●	

S10

Synthesis and analysis of an aluminium oxalato compound

In this experiment an aluminium oxalato compound is analysed and its chemical formula determined.

Equipment

- 2 x 250 cm^3 beaker
- spatula
- weighing bottle and lid
- 50 cm^3 measuring cylinder
- 500 cm^3 conical flask
- 20 cm^3 pipette
- 50 cm^3 burette
- 2 x 250 cm^3 conical flask
- water pump
- Buchner funnel
- Buchner flask
- filter paper
- tripod and gauze
- Bunsen burner
- glass rod
- watch glass
- ice-water bath
- funnel

Reagents

- aluminium turnings
- potassium hydroxide
- oxalic acid dihydrate
- methylated spirit
- sodium oxalate
- potassium manganate(VII) solution
- dilute sulfuric acid

LIVERPOOL
JOHN MOORES UNIVERSITY
AVRIL ROBARTS LRC
TITHEBARN STREET
LIVERPOOL L2 2ER
TEL. 0151 231 4022

RS•C

Safety

Potassium hydroxide: Corrosive. Burns skin, eyes and other tissues. Harmful if ingested.

Oxalic acid dihydrate: Dust irritates respiratory system. Harmful in contact with skin or if swallowed. Dust and solutions irritate eyes.

Methylated spirit: Highly flammable. Toxic by inhalation and if swallowed.

Sodium oxalate: Harmful if ingested or inhaled. Irritating to skin and eyes.

Potassium manganate(VII) solution: Extremely destructive to skin, eyes, upper respiratory tract and mucous membranes. Harmful if inhaled, ingested or absorbed through skin. Inhalation may be fatal.

Dilute sulfuric acid: Irritating to skin and eyes. May cause burns.

Gloves should be worn while handling oxalic acid, potassium hydroxide and potassium manganate(VII) solution.

Synthesis of oxalato complex

Weigh out 1 g of aluminium turnings into a 250 cm^3 beaker. After making sure there are no naked flames nearby, add a solution of potassium hydroxide made by dissolving 6 g of potassium hydroxide in 50 cm^3 of distilled water.

Which gas is evolved? Write a balanced equation for the reaction.

Stir the solution occasionally and, when the effervescence has moderated, place the beaker on a tripod and gauze, and heat it carefully until the solution begins to boil gently. Cover the boiling solution with a watch glass. The aluminium should soon dissolve, leaving a fine residue in suspension. Filter the hot solution into a 500 cm^3 conical flask at a water pump, and pour the filtrate into another clean, dry 250 cm^3 beaker. Swill the inside of the flask with 10 cm^3 of distilled water, and add this to the solution in the beaker, thus minimising the loss of solution during the filtration process.

Place the beaker on a tripod and gauze, and heat carefully until the solution boils again. In the meantime, weigh out 14 g of oxalic acid dihydrate. Add the oxalic acid to the gently boiling solution in about six approximately equal portions. Stir the solution immediately after the addition of each portion. When all the oxalic acid has been added, boil the solution gently for a short time until there is very little or no solid suspension, but do not let the solution get too concentrated. If necessary, filter off any suspension as above, again transferring the filtrate into a clean 250 cm^3 beaker and rinsing out the filtration flask with 10 cm^3 of distilled water.

Allow the filtrate to cool to room temperature.

Steadily add about 50 cm^3 of methylated spirit to the solution in the beaker, stirring constantly during the addition. Stand the beaker in the centre of a crystallising dish, and build up an ice and cold water mush around it until the mush is higher than the surface of the liquid in the beaker. Allow the solution to cool in this way for 15 minutes, stirring frequently to loosen any precipitate that adheres to the inside of the beaker; leave it in the cooling bath for a further 5 minutes without stirring.

Filter off the crystals and wash with an ice-cold mixture of equal volumes of methylated spirit and distilled water. Wash twice more with methylated spirit (2 x 10 cm^3). Transfer the crystals to a watch glass, and break them up with a spatula. Allow to dry.

Analysis of the complex

In this part of the experiment the chemical formula of the complex is determined by quantitative analysis. Some of the information required for this process is provided.

It can be shown that the compound contains groups of two carbon atoms and four oxygen atoms, which behave in many reactions like the oxalate anion, $C_2O_4^{2-}$. In this section a reaction is carried out in which a solution of the compound behaves in exactly the same way as a solution of sodium oxalate, $Na_2C_2O_4$. The mass of the compound which contains one mole of oxalate must be determined.

To carry out the analysis a solution of potassium manganate(VII) of accurately known concentration is needed. The solution of sodium oxalate which you prepare is a primary standard, and reacts with potassium manganate(VII) in a simple molar ratio between 60–100 °C. It can therefore be used to determine the concentration of the manganate solution. Once standardised, the manganate solution can be used to determine the percentage of oxalate in the compound you have prepared.

Weigh accurately about 0.15 g of sodium oxalate into a 250 cm^3 conical flask. Repeat using a second flask, taking care to note which flask contains which mass of oxalate.

To each of the conical flasks, add about 50 cm^3 of dilute sulfuric acid, and then rinse the inside of the conical flask with a little distilled water so that all of the solid sodium oxalate is washed down into the acid.

Place around 150 cm^3 of potassium manganate(VII) solution in a beaker. Keep the beaker covered with a watch glass when it is not in use. Fill a burette with the potassium manganate(VII) solution. Heat one of the oxalate flasks gently until it boils.

Remove the flask from the tripod, and stand it on a piece of white paper beneath the burette. Titrate with the manganate solution, until one drop leaves the solution in the flask with a pink tinge. If at any time a permanent brown precipitate appears in the solution, reheat the flask and continue the titration. Repeat the titration procedure with the second flask of oxalate.

The equation for the reaction is

$$2MnO_4^-{}_{(aq)} + 5C_2O_4^{2-}{}_{(aq)} + 16H^+{}_{(aq)} \rightleftharpoons 2Mn^{2+}{}_{(aq)} + 10CO_2{}_{(g)} + 8H_2O{}_{(l)}$$

On the basis of this information, calculate the concentration of the manganate solution.

Weigh accurately about 0.2 g of your aluminium compound into a clean 250 cm^3 conical flask. Repeat using a second flask.

Repeat the titration procedure above, beginning at the point where 50 cm^3 of dilute sulfuric acid was added to each flask.

Calculate the mass of aluminium compound that contains one mole of oxalate.

Besides oxalate the compound also contains potassium and aluminium. It has been found that:

      460 g of the compound contains 1 mole of aluminium

      153 g of the compound contains 1 mole of potassium

Does the sum of the aluminium, potassium and oxalate content account for the total composition of the compound? If not, what is the source of the discrepancy?

These ideas can be confirmed by heating some of the compound gently in a test-tube, or by taking an infrared spectrum of the compound in a Nujol mull, and comparing with the spectrum of sodium oxalate.

Synthesis and analysis of an aluminium oxalato compound

In this activity an aluminium oxalato compound is synthesised, and analysed to determine its chemical formula.

Synthesis of the complex

The compound seems to become partially dehydrated by the methylated spirit, and needs some time in the atmosphere to pick up water before the ideal formula, $K_3Al(C_2O_4)_3.3H_2O$, is attained.

$$OH^-_{(aq)} + 3H_2O_{(l)} + Al_{(s)} \rightleftharpoons [Al(OH)_4]^-_{(aq)} + \frac{3}{2}H_{2\ (g)}$$

Analysis of the complex

Students determine the oxalate content of the compound. They are given other experimental results, which tell them the mass of compound containing one mole of potassium, and the mass of compound containing one mole of aluminium. From these, they calculate the ratio of the number of moles of potassium, aluminium and oxalate in the compound.

By heating some of the compound in a clean dry test-tube, students can identify water as the missing constituent. Comparing the infrared spectrum of the compound with the sodium oxalate spectrum should reveal the O–H stretch in the spectrum of the compound only.

RS•C

Synthesis and analysis of an aluminium oxalato compound

Equipment

- 2 x 250 cm^3 beaker
- weighing bottle and lid
- 50 cm^3 measuring cylinder
- 500 cm^3 conical flask
- 20 cm^3 pipette
- 50 cm^3 burette
- 2 x 250 cm^3 conical flask
- water pump
- Buchner funnel
- Buchner flask
- filter paper
- tripod and gauze
- Bunsen burner
- glass rod
- watch glass
- ice-water bath
- funnel

Reagents

- aluminium turnings
- potassium hydroxide
- oxalic acid
- methylated spirit
- sodium oxalate
- potassium manganate(VII) solution
- dilute sulfuric acid

11. Catalase investigation

This is an open-ended activity investigating various factors that affect the reaction between catalase and hydrogen peroxide. A number of suggestions for different factors that could be investigated are given, however very little explicit information about experimental procedure is provided.

This experiment is suitable for a wide range of ability levels, as students can approach it at their own level. It also lends itself particularly well to group work. The work could be carried out in a single session, or extended over a number of sessions depending on the needs of the students involved, and the time available.

This experiment is suitable for:

■ students with a wide range of abilities

■ a single laboratory session, with the possibility of extension

■ group work

Activity type

| formal | ○ | experimental | ○ | divergent | ○ | investigatory | ● |

Skills

manipulation	weighing	●	volume	●	handling	●						
techniques	reflux/distillation	○	recrystallisation	○	chromatography	○	inert atmosphere	○	spectroscopy	○	titration	○
observation	colour	○	volume	●	temperature	○	pressure	○	physical state	○		
data collection	qualitative	○	numerical	●	spectral	○	electronic	○				
data processing	calculation	●	computing	○	matching	●						
interpretation	selection	○	validation	●	deduction	●	prediction	○				
problem solving	identification	○	in/output	○	breakdown	○	methods	●	assembly	●		
team work	skills identification	○	analysis	○	role assignment	○	organisation	○	resources	○	outcomes	○
experiment design	input	○	output	○	precision	●	techniques	●	validity	●		
communication	report	●	poster	○	oral	○	audience	○				
safety	COSHH	○	application	●	review	○	disposal	○				

S11

Catalase investigation

In this activity, the reaction between the enzyme catalase and hydrogen peroxide is studied. Catalase is extracted from natural sources, the optimum conditions of the reaction are investigated, and the effect of various inhibitors evaluated.

At the end of the activity, you should be able to:

■ plan a scientific investigation;

■ recognise the different variables that require control;

■ communicate your plans, rationale and results clearly; and

■ suggest reasons for your observations, and test these if possible.

The two enzymes catalase and superoxide dismutase protect cells of aerobic organisms against attack by the superoxide radical anion ($O_2^{\bar{\bullet}}$) and hydrogen peroxide (H_2O_2). $O_2^{\bar{\bullet}}$ is a harmful by-product of the metabolic oxidation of fats and carbohydrates. Superoxide dismutase converts superoxide ions to hydrogen peroxide, which is then converted to water and oxygen by catalase.

$$2O_2^{\bar{\bullet}} + 2H^+ \xrightarrow{\text{superoxide dismutase}} H_2O_2 + O_2$$

$$2H_2O_2 \xrightarrow{\text{catalase}} 2H_2O + O_2$$

Although the precise structure of catalase varies between organisms, its general quaternary structure is analogous to that of haemoglobin. It is tetrameric, and each sub-unit contains an iron-centred porphyrin ring. However, haemoglobin contains iron in the Fe(II) oxidation state, while catalase contains iron in the Fe(III) state. This iron-centred porphyrin ring is believed to be the active site of catalase.

Catalase is gradually, irreversibly oxidised by hydrogen peroxide, so at low enzyme concentrations or over long reaction periods (of more than about 3 minutes) there can be significant deviation from first order behaviour.

There are many ways of following the progress of the reaction of catalase and hydrogen peroxide, including measuring the rate of oxygen evolution, measuring the heat production of the reaction, and determining the concentration of hydrogen peroxide by spectrometry or titration. In this experiment the rate of oxygen evolution is measured.

Equipment

- food blender
- cheesecloth (cut into 30 cm squares)
- centrifuge (optional)
- test-tubes with side arms
- stoppers for test-tubes
- rubber tubing (to fit side arm of test-tubes)
- graduated cylinders (various sizes)
- 1000 cm^3 beaker
- dropping pipettes
- thermometer
- water bath
- stop-clock

Reagents

- hydrogen peroxide solution (3%)
- phosphate buffer solution and/or deionised water
- universal indicator
- hydrochloric acid solution (0.5 mol dm^{-3})
- sodium hydroxide solution (0.5 mol dm^{-3})

Inhibitors
- ethanol
- methanol
- ethanoic acid

Different substrate or catalyst
- copper(II) chloride solution (0.1 mol dm^{-3})
- *tert*-butylhydroperoxide solution (3%)

Safety

Hydrogen peroxide (3%): Irritant to skin, eyes and mucous membranes.

Ethanol: Harmful by inhalation or ingestion. Highly flammable.

Methanol: Highly flammable. Toxic by ingestion, inhalation and skin contact. Can cause delayed damage to eyes if ingested.

Ethanoic acid: Flammable. Irritant to eyes and skin.

***tert*-Butyl hydroperoxide (3%):** Irritant to skin and eyes.

Direct contact with all of the above reagents should be avoided. In the case of contact with skin or eyes wash with copious amounts of water and notify a demonstrator. Gloves should be worn whilst handling *tert*-butyl hydroperoxide and ethanoic acid.

Procedure

Design an experiment to investigate the reaction rate of catalase and hydrogen peroxide.

Prepare catalase extracts by combining plant or animal material with deionised water or phosphate buffer solution, in a blender. The extract will probably need filtering through cheesecloth. More consistent results may be obtained if the extracts are centrifuged and the resulting supernatant liquid used.

A good starting point is to use material extracted from potatoes. Follow the reaction by adding about 3 cm^3 of the catalase extract to about 5 cm^3 of 3% hydrogen peroxide solution. The amount of plant or animal material used may have to be altered to get a measurable reaction rate.

Further work can be chosen from the topics outlined below. Make sure that investigations are fully planned before starting them. Attempt to rationalise your observations; perform further experiments if necessary. Consult a demonstrator if the equipment required has not been provided. There may be another way to carry out the planned work using available resources, or it may be possible to obtain the required materials.

Sources of catalase
What sources of catalase give the highest activity? Measure the relative rates of activity for catalase from at least three different sources.

pH
What effect does pH have on the activity of catalase? Does catalase permanently denature at extremes of pH?

Temperature
What effect does temperature have on the activity of catalase? Does catalase permanently denature at extremes of temperature?

Inhibition
How do the various reagents supplied affect the rate of reaction?

Change of substrate or catalyst
A number of reagents have been provided, which may either catalyse the decomposition of hydrogen peroxide, or be decomposed by catalase. Measure the relative reaction rates using at least three of these, in place of either hydrogen peroxide or catalase.

Other methods of following the reaction
Plan an experiment using another method to follow the reaction between catalase and hydrogen peroxide. Comment on its effectiveness.

Stability of catalase extracts

Devise an experiment to test how long the catalase extracts retain their activity. Identify the variables.

Catalase is an extremely efficient enzyme. Reaction rates of catalase and hydrogen peroxide approach the diffusion controlled limit. What does the term 'diffusion controlled limit' mean?

Catalase investigation

This relatively simple experiment is ideal for allowing students to work in an open-ended manner. It involves the preparation of a catalase extract, followed by measurements of the rate of oxygen evolution when various solutions are heated and mixed. The preparation of gas collection apparatus uses common equipment, and is simple enough to allow students to devise their own method. General instructions for the preparation of catalase extracts are given in the student guide, but the details remain to be determined by the students themselves.

Procedure

Students are given the task of preparing catalase extracts from a variety of plant or animal sources. This can be done very simply by combining equal masses of plant or animal material with deionised water or phosphate buffer solution in a blender. The extracts may need filtering through cheesecloth. More consistent results are obtained if extracts are centrifuged and the resulting supernatant liquid used.

A simple gas collection apparatus can be assembled using a stoppered test-tube with a side arm, connected by rubber tubing to a graduated cylinder that has been filled with water and inverted in a large water-filled beaker. This allows students to easily monitor the volume of oxygen generated by the reaction as a function of time.

The reaction can also be analysed using UV/visible spectroscopy. The rate of disappearance of hydrogen peroxide can be followed by observing the absorbance at $\lambda = 240$ nm. This corresponds to a transition in hydrogen peroxide.

Discussion

Sources of catalase
It is important that students use the same mass of each source of catalase, in order to make a valid comparison. This has deliberately not been stated in the student guide, and may have to be suggested to some of them.

pH
Catalase activity should decrease by half when the pH is lowered from 7 to 3. Acetate and formate are not recommended for pH studies as they have inhibitory effects. Above a pH of about 10, the catalase will denature.

Temperature
The optimum temperature for catalase activity is approximately 40 °C, above which the enzyme starts to denature.

Inhibition
There are many substances that inhibit the function of catalase, of which a large proportion are highly toxic. Relatively safe inhibitors include acetate, ascorbate, ethanol, formate, methanol and nitrite. Only small amounts of these inhibitors should be needed to significantly slow the rate of reaction.

Different substrate or catalyst

$CuCl_2$ can catalyse the decomposition of hydrogen peroxide. Hydrogen peroxide can be substituted with *tert*-butylhydroperoxide. The relative rates of reaction using both of these can be compared to the rate of reaction between catalase and hydrogen peroxide.

Other methods of following reaction

Some other methods include calorimetry, titration to determine the amount of residual hydrogen peroxide, or electrochemical analysis. Further details of techniques can be found in H U Bergmeyer, *Methods of enzymatic analysis*, VCH Verlagsgesellschaft, Weinheim, 1986.

Stability of catalase extracts

Storage temperature or pH of the solution may affect how long the catalase extracts keep their activity, as well as whether they are stored in an airtight container, or have been made with phosphate buffer or distilled water.

Catalase is an extremely efficient enzyme. Reaction rates of catalase and hydrogen peroxide approach the diffusion controlled limit. The defintion of a diffusion controlled reaction is a reaction whose rate is determined by the rate at which molecules diffuse through the solvent.

Reference

The experiment is adapted from D R Kimbrough, M A Magoun and M Langfur, *J. Chem. Educ.*, 1997, **74**, 210.

Catalase investigation

Equipment

- food blender
- cheesecloth (cut into 30 cm squares)
- centrifuge (optional)
- test-tubes with side arms
- stoppers for test-tubes
- rubber tubing (to fit side arm of test-tubes)
- graduated cylinders (various sizes)
- 1000 cm^3 beakers
- dropping pipettes
- thermometers
- water baths
- stop-clock

Reagents

- hydrogen peroxide solution (3%)
- phosphate buffer solution and/or deionised water
- universal indicator
- hydrochloric acid solution (0.5 mol dm^{-3})
- sodium hydroxide solution (0.5 mol dm^{-3})

Inhibitors
- ethanol
- methanol
- ethanoic acid

Different substrate or catalyst
- copper(II) chloride solution (0.1 mol dm^{-3})
- *tert*-butylhydroperoxide solution (3%)

A number of different plant or animal materials from which catalase can be extracted must also be supplied.

12. Propanone iodination

In this activity the kinetics of propanone iodination are investigated. The order of reaction with respect to each of the reactants, and the overall rate of reaction, are determined.

This experiment is suitable for:

■ first or second year students

■ approximately four hours

■ group or individual work

Activity type

| formal | ○ | experimental | ● | divergent | ○ | investigatory | ○ |

Skills

manipulation	weighing	○	volume	○	handling	●						
techniques	reflux/distillation	○	recrystallisation	○	chromatography	○	inert atmosphere	○	spectroscopy	○	titration	○
observation	colour	●	volume	●	temperature	○	pressure	○	physical state	○		
data collection	qualitative	○	numerical	●	spectral	○	electronic	○				
data processing	calculation	●	computing	○	matching	●						
interpretation	selection	○	validation	○	deduction	●	prediction	○				
problem solving	identification	○	in/output	○	breakdown	○	methods	○	assembly	○		
team work	skills identification	○	analysis	○	role assignment	○	organisation	○	resources	○	outcomes	○
experiment design	input	○	output	○	precision	○	techniques	○	validity	○		
communication	report	●	poster	○	oral	○	audience	○				
safety	COSHH	○	application	●	review	○	disposal	○				

S12

Propanone iodination

In this activity, the order of reaction with respect to each of the reactants involved in the iodination of propanone in acidic solution is determined. The rate of reaction is measured by the removal of aliquots from the reaction at regular intervals, quenching the reaction, and titrating the unreacted iodine against sodium thiosulfate.

Equipment

■ 6 x 250 cm^3 conical flasks

■ 100 cm^3 conical flask

■ water bath with thermostat

■ 50 cm^3 graduated pipette

■ pipette filler

■ 20 cm^3 pipette

■ 50 cm^3 burette

■ dropping pipette

■ stop-clock

Reagents

■ propanone

■ sulfuric acid (1 mol dm^{-3})

■ iodine solution (0.1 mol dm^{-3}) in KI solution (0.2 mol dm^{-3})

■ sodium ethanoate solution (1 mol dm^{-3})

■ sodium thiosulfate solution (0.01 mol dm^{-3})

■ starch solution

Safety

Propanone: Highly flammable.

Sulfuric acid: Irritating to skin and eyes. Causes burns.

Procedure

Make up each of the solutions in the table below in stoppered 250 cm^3 conical flasks. Place them, along with a flask containing about 120 cm^3 of 0.1 mol dm^{-3} iodine solution, in a water bath with a thermostat set at 25 °C. Leave the solutions for around twenty minutes to allow time for the temperature to equilibrate.

Add 20 cm^3 of iodine solution to the first flask, and swirl to mix. Start a stop-clock immediately after adding the iodine. After five minutes withdraw a 20 cm^3 sample of the solution, and add to it about 10 cm^3 of 1 mol dm^{-3} sodium ethanoate solution. Then titrate the remaining iodine against 0.01 mol dm^{-3} sodium thiosulfate solution, using a few drops of starch near the end point as an indicator. Repeat every five minutes until no reaction solution remains.

Repeat this procedure for all of the solutions.

Run	Propanone/cm^3	1 mol dm^{-3} H$_2$SO$_4$/cm^3	H$_2$O/cm^3
1	20	10	150
2	15	10	155
3	10	10	160
4	20	15	145
5	20	5	155

For each run, plot a graph of the concentration of iodine against time. Determine the order of the reaction with respect to iodine after discussing your results with a demonstrator.

Use your experimental data to find out how k' varies with the concentration of propanone, and hence the order of the reaction with respect to propanone. Do the same for H$^+$.

On the basis of your results, suggest which species are involved in the rate determining step.

Propanone iodination

In this activity, students are expected to determine the order of reaction with respect to each of the reactants involved in the iodination of propanone in acidic solution. The rate of reaction will be determined by removing aliquots from the reaction at regular intervals, quenching the reaction, and titrating the unreacted iodine against sodium thiosulfate.

Pre-laboratory work

It is recommended that a pre-laboratory session is carried out in the form of a guided discussion, preferably with groups of around twelve students. A set of notes is provided in order to give an idea of the direction such a discussion could follow. Laboratory schedules should not be handed to students until completion of the pre-laboratory session.

The reaction under investigation is

$$CH_3COCH_3 + I_2 \longrightarrow CH_2ICOCH_3 + H^+ + I^-$$

Explain briefly how the mechanisms proposed for a reaction should be compatible with the rate equation.

Students may know, or be able to work out, equations (1), (2) and (3). It is worth asking them first, and guiding them through working this out. They may need reminding that the orders a, b, c ... are not related to the stoichiometry of the reaction (a common misconception), and can only be determined experimentally.

$$\text{rate} = k[A]^a[B]^b[C]^c... \tag{1}$$

The rate of reaction can be measured by finding out how the concentration of A varies with time. If the concentrations of B, C... are reasonably constant, then equation (1) can be written;

$$\text{rate} = -\frac{d[A]}{dt} \simeq k'[A]^a \tag{2}$$

where k' is the effective rate coefficient;

$$k' = k[B]^b[C]^c... \tag{3}$$

By integrating equation (2), we can obtain;

$$[A] \simeq [A]_0 - k't \qquad \text{for } a = 0 \tag{4}$$

$$\ln [A] \simeq \ln [A]_0 - k't \qquad \text{for } a = 1 \tag{5}$$

$$\frac{1}{[A]} \simeq \frac{1}{[A]_0} + k't \qquad \text{for } a = 2 \tag{6}$$

where $[A]_0$ is the concentration of A at the start of the reaction ($t = 0$).

It is not advisable to go into the detailed mathematics of these integrations, but simply to present them to the students as true, unless they are particularly interested in the mathematics. Students must understand the use of graphs of [A], ln[A], and 1/[A], against time to find out the order of a reaction with respect to a particular reactant.

In order to write a full rate equation, the orders b, c... with respect to the other possible reactants B, C... must be determined. To do this the most obvious method would be to observe the rate of change of [B], [C]... and use equations analogous to (4), (5) and (6). However, as this is not possible for many reactions, the initial concentrations of $[B]_0$, $[C]_0$... are varied and the rate of $-d[A]/dt$ is measured instead. The effective rate constant, k', can be found using equations (4), (5) or (6). From equation (3) we obtain the equation below:

$$\ln k' = \ln k + b \ln [B]_0 + c \ln [C]_0 + \dots \qquad (7)$$

Therefore, a graph of $\ln k'$ against $\ln[B]_0$ at constant $[A]_0$, $[C]_0$... will have a slope b. The order of reaction with respect to the other reactants can be obtained in an analogous manner.

Students should now understand that they need to measure the rate of disappearance of one of the reactants, [A] (iodine in this case), with respect to time, while the concentrations of the other reactants remain constant. The order of the reaction with respect to A can be determined by plotting graphs of [A], ln[A], and 1/[A], against time. The effective rate coefficient, k', can be determined from the slope of the appropriate graph. By then varying the initial concentrations of each of the other reactants in turn, the order with respect to each of them can be determined using equation (7) as detailed above.

Students will need some help in planning how this could actually be done. Reminding them of the reaction between iodine and sodium thiosulfate may prompt them into thinking of removing aliquots of the reaction mixture at regular intervals, and titrating excess iodine to determine the rate of consumption of iodine. They will probably need to be told that the reaction can be quenched using sodium ethanoate, and that the reaction rate should be measured at constant temperature, *ie* in a water bath.

Before handing out experimental schedules, make sure that the students have decided how much of each reactant they would want for each run, to ensure that only one of the reactants' initial concentrations is changed at a time. Students should now be ready to read the text, and start organising how they are going to carry out the experiment.

Propanone iodination

Equipment

- 6 x 250 cm^3 conical flasks
- 100 cm^3 conical flask
- water bath with a thermostat
- 50 cm^3 graduated pipette
- pipette filler
- 20 cm^3 pipette
- 50 cm^3 burette
- dropping pipette
- stop-clock

Reagents

- propanone
- sulfuric acid (1 mol dm^{-3})
- iodine solution (0.1 mol dm^{-3})
- sodium ethanoate solution (1 mol dm^{-3})
- sodium thiosulfate solution (0.01 mol dm^{-3})
- starch solution

13. An unknown metal

In this activity students work through a series of experiments testing the solubilities of various salts of an unknown metal, M. These results are then compared with the solubilities of known salts.

Observation is the most important part of this experiment. It is suitable for first year students, and can be tackled individually or in groups.

This experiment is suitable for:

■ first year students

■ 2 hours

■ group or individual work

Activity type

formal	○	experimental	●	divergent	○	investigatory	○

Skills

manipulation	weighing	○	volume	○	handling	●						
techniques	reflux/distillation	○	recrystallisation	○	chromatography	○	inert atmosphere	○	spectroscopy	○	titration	○
observation	colour	●	volume	○	temperature	○	pressure	○	physical state ●			
data collection	qualitative	●	numerical	○	spectral	○	electronic	○				
data processing	calculation	○	computing	○	matching	●						
interpretation	selection	○	validation	○	deduction	●	prediction	●				
problem solving	identification	●	in/output	○	breakdown	○	methods	○	assembly	●		
team work	skills identification	○	analysis	○	role assignment	○	organisation	○	resources	○	outcomes	○
experiment design	input	○	output	○	precision	○	techniques	○	validity	○		
communication	report	●	poster	○	oral	○	audience	○				
safety	COSHH	○	application	●	review	○	disposal	○				

An unknown metal

This experiment involves observation more than reasoning or manipulative skills. A series of experiments which tests the solubilities of various salts of an unknown metal, M, are undertaken and the results compared with solubilities of known salts.

Equipment

- dropping pipettes
- test-tubes
- test-tube holder
- test-tube rack
- boiling-tube
- 25 cm^3 measuring cylinder
- hot plate
- Bunsen burner

Reagents

- unknown metal M sulfate solution (0.1 mol dm^{-3})
- aluminium sulfate solution (0.1 mol dm^{-3})
- barium nitrate solution (0.1 mol dm^{-3})
- lead nitrate solution (0.1 mol dm^{-3})
- silver nitrate solution (0.1 mol dm^{-3})
- potassium nitrate
- hydrochloric acid (2 mol dm^{-3})
- ammonia solution (2 mol dm^{-3})
- potassium bromide solution (0.1 mol dm^{-3})
- potassium iodide solution (0.1 mol dm^{-3})
- sodium carbonate solution (0.1 mol dm^{-3})
- sodium hydroxide solution (0.1 mol dm^{-3})
- sulfuric acid (2 mol dm^{-3})
- sodium tetraphenylboron solution (0.1 mol dm^{-3})
- aluminium sulfate
- 8-hydroxyquinoline
- ethanoic acid (2 mol dm^{-3})
- chlorine gas

Safety

M compounds: Very poisonous if taken internally.

Aluminium sulfate: Irritant to skin, eyes and respiratory system.

Barium nitrate: Contact with combustible material may cause fire. Harmful by inhalation and if swallowed.

Lead nitrate: Harmful by inhalation and if swallowed.

Silver nitrate: Irritates eyes and causes burns.

Potassium nitrate: Irritating to skin, eyes and respiratory system.

Hydrochloric acid: Irritating to eyes, skin and respiratory system. Causes burns.

Ammonia solution: Irritating to skin, eyes and respiratory system. Causes burns.

Potassium bromide: Harmful if ingested in quantity. Irritating to eyes. May evolve toxic fumes in fire.

Potassium iodide: May cause sensitisation by inhalation and skin contact. Possible risk of harm to unborn child. Irritating to eyes, skin and respiratory system.

Sodium carbonate: Irritating to eyes, skin and respiratory system.

Sodium hydroxide: Irritating to skin. Severely irritating to eyes.

Sulfuric acid: Irritating to skin and eyes. Causes burns.

8-Hydroxyquinoline: Irritating to skin, eyes, respiratory system and mucous membranes. Possible carcinogen.

Ethanoic acid: Flammable. Irritant to eyes and skin.

Chlorine gas: May be fatal if inhaled.

Procedure

Solutions of M sulfate, aluminium sulfate, barium nitrate, lead nitrate and silver nitrate are provided. Test the solubilities of the chlorides, bromides, iodides, sulfates, carbonates, hydroxides, tetraphenylboron compounds and oxine complexes of the cations by adding a 1 cm depth of the reagents listed below to 1 cm depths of the cation solutions in test-tubes. In some cases additional instructions are given. Record observations in the table provided. If no precipitate forms put down a cross, and record the colour of any precipitates that do form. In the two cases where a potassium solution is tested, dissolve a little potassium nitrate in water.

Chloride

Add dilute hydrochloric acid to the cation solutions. Some of the precipitates have unusual properties. Gently heat the test-tube containing lead nitrate to boiling point, and then cool under a tap. Repeat with the test-tube containing your unknown cation. Add excess ammonia solution to the test-tube containing silver cations. Repeat with the cooled test-tube containing the unknown cation.

Bromide

Add potassium bromide solution. Add excess ammonia to the silver solution.

Iodide

Add potassium iodide solution. Add excess ammonia to the silver solution.

Carbonate

Add sodium carbonate solution.

Hydroxide

Add sodium hydroxide solution a drop at a time at first, shaking gently after each drop.

Sulfate

Add dilute sulfuric acid.

Tetraphenylboron compounds

Add sodium tetraphenylboron solution, $NaB(C_6H_5)_4$. In the case of aluminium, make up some fresh aluminium sulfate solution by just filling the rounded bottom of a test-tube with solid aluminium sulfate and adding a 3 cm depth of distilled water. (The need to prepare this solution arises because bench solutions of aluminium sulfate may be contaminated with potassium.)

Oxine

Place some 8-hydroxyquinoline crystals in a dry test-tube and add a 5–6 cm depth of 2 mol dm^{-3} ethanoic acid. Stir vigorously until the solid dissolves. Make this solution as concentrated as possible. Add 2 mol dm^{-3} ammonia solution dropwise with shaking until the precipitate that forms after each drop just fails to redissolve. Then add 2 mol dm^{-3} ethanoic acid drop by drop until the precipitate just dissolves.

Add the prepared solution to the metal ion solution, reserving some for the next section.

Using the table of results, try to answer the following questions.

1   M definitely does not behave in the same way as two of the cations. Which are they?

2   When is the solubility or insolubility of a particular salt of the M cation especially striking, in the sense that it adds a new exception to the generalisations made in the Appendix?

3   In its departures from the generalisations made in the Appendix, does the M cation resemble any of the cations with which it is being compared? State why. There are three such cations.

4   Salts of the M cation give a brightly coloured spectrum when placed in the Bunsen flame. When analysed, this spectrum is very simple, like those of the alkali metal cations. Does this parallel part of your conclusions from **3**?

Other compounds of the metal

Carry out the following experiments, recording any observations for (b)–(g) in the same results table.

(a) Pour 5 cm^3 of the solution of the M cation into a clean boiling-tube and acidify with dilute hydrochloric acid until no more chloride precipitates. Then bubble chlorine gas from a cylinder in the fume cupboard into the solution; it will clear within a minute.

What has happened to the unknown metal? Continue with the remainder of the experiments even if the answer to this question is not clear, and try to answer the question again when the experiment is complete.

Boil off the dissolved chlorine in another fume cupboard, cool the boiling-tube under the tap, and use 1 cm depths of this solution for each of the tests (b)–(g) below, in separate test-tubes:

(b) Add potassium bromide solution.

(c) Add sodium carbonate solution.

(d) Add dilute sulfuric acid.

(e) Add dilute sodium hydroxide solution.

(f) Add dilute ammonia solution drop by drop until a permanent precipitate just appears. Then add dilute ethanoic acid until the solution just becomes clear. Test this solution with the remainder of the 8-hydroxyquinoline reagent.

(g) To a 1 cm depth of the solution, add 1 cm depth of iron(II) sulfate solution. Gently heat to boiling, and then cool under running tap-water.

Explain any observations made. Do they confirm your answer, or enable you to answer the question in (a)?

Based on knowledge of the behaviour of their cations, could element M be an alkali metal or alkaline earth metal?

To which group of the Periodic Table might the unknown metal belong? Discuss any suggestions with a demonstrator.

Appendix

Anion	Generalisation	Exceptions
ethanoates	all soluble	Ag^+
nitrates	all soluble	none
perchlorates	all soluble	K^+, Rb^+, Cs^+, NH_4^+
chlorides	all soluble	$Ag^+, Pb^{2+}, Hg^{2+}, Cu^+$
bromides	all soluble	$Ag^+, Pb^{2+}, Hg^{2+}, Cu^+$
iodides	all soluble	$Ag^+, Pb^{2+}, Hg^{2+}, Cu^+$
sulfates	all soluble	$Ca^{2+}, Sr^{2+}, Ba^{2+}, Pb^{2+}, Rb^+$
tetraphenylboron compounds	all soluble	$K^+, Rb^+, Cs^+, NH_4^+, Ag^+$
hydroxides	all insoluble	alkali metal cations, $H^+, NH_4^+, Sr^{2+}, Ba^{2+}, Ra^{2+}$
phosphates	all insoluble	$Na^+, K^+, Rb^+, Cs^+, H^+, NH_4^+$
carbonates	all insoluble	$Na^+, K^+, Rb^+, Cs^+, H^+, NH_4^+$
sulfites	all insoluble	$Na^+, K^+, Rb^+, Cs^+, H^+, NH_4^+$
sulfides	all insoluble	alkali metal cations, H^+, NH_4^+, alkaline earth metal cations

Table for recording observations of the solubilities of the salts of metal M and various others in aqueous solution

Reagent	M	Aluminium	Barium	Lead	Silver	Potassium	Cation M + Cl$_2$ (g)
chloride			×			×	
bromide			×	off-white precipitate		×	
iodide			×			×	—
carbonate		white precipitate	white precipitate			×	
hydroxide				white precipitate	brown precipitate	×	
sulfate	×	×		white precipitate	×	×	
tetraphenylboron			×	×			—
oxine		precipitate in yellow solution	×	×			

An unknown metal

This experiment involves observation more than reasoning or manipulative skills. A series of experiments which test the solubilities of various salts of an unknown metal, M, are undertaken and the results compared with solubilities of known salts. The metal, M, is thallium.

In the first part, students may realise that M behaves like:

■ Ag^+ in the insolubility and appearance of its chloride, bromide and iodide;

■ Pb^{2+} in the insolubility and appearance of its chloride, bromide, and iodide, and in the solubility of its chloride in hot water; and

■ K^+ in the solubility of its hydroxide and carbonate, in the insolubility of its tetraphenylboron compound, and in its spectral characteristics.

In the second part, students examine compounds of thallium in higher oxidation states. They oxidise TlCl to Tl(III) by passing chlorine through a suspension of the insoluble chloride in dilute HCl. They should see that the oxidised cation behaves like Al^{3+}, demonstrating that the properties of the state of maximum valency display chemical periodicity most clearly.

The reduction, part (g), of Tl(III) with iron(II) sulfate is carried out in order to confirm the answer to the question in part (a). On cooling, a precipitate of TlCl forms in the brown Fe(III) solution.

Table for recording observations of the solubilities of the salts of metal M and various others in aqueous solution

Reagent	M	Aluminium	Barium	Lead	Silver	Potassium	Cation M + Cl$_{2(g)}$
chloride	white	×	×	white	white	×	×
bromide	white	×	×	off-white precipitate	cream	×	×
iodide	yellow	×	×	yellow	yellow	×	—
carbonate	×	white precipitate	white precipitate	white	brown	×	brown
hydroxide	×	white	×	white precipitate	brown precipitate	×	brown
sulfate	×	×	white	white precipitate	×	×	×
tetraphenylboron	white	×	×	×	white	precipitate	—
oxine	×	precipitate in yellow solution	×	×	white	precipitate	yellow

An unknown metal

Equipment

- dropping pipettes
- test-tubes
- test-tube holder
- test-tube rack
- boiling-tube
- 25 cm^3 measuring cylinder
- hot plate
- Bunsen burner

Reagents

- thallium(I) sulfate solution (0.1 mol dm^{-3}) – unknown metal solution
- aluminium sulfate solution (0.1 mol dm^{-3})
- barium nitrate solution (0.1 mol dm^{-3})
- lead nitrate solution (0.1 mol dm^{-3})
- silver nitrate solution (0.1 mol dm^{-3})
- potassium nitrate
- hydrochloric acid (2 mol dm^{-3})
- ammonia solution (2 mol dm^{-3})
- potassium bromide solution (0.1 mol dm^{-3})
- potassium iodide solution (0.1 mol dm^{-3})
- sodium carbonate solution (0.1 mol dm^{-3})
- sodium hydroxide solution (0.1 mol dm^{-3})
- sulfuric acid (2 mol dm^{-3})
- sodium tetraphenylboron solution (0.1 mol dm^{-3})
- aluminium sulfate
- 8-hydroxyquinoline
- ethanoic acid (2 mol dm^{-3})
- chlorine gas

PROGRESSIVE
DEVELOPMENT
OF PRACTICAL
SKILLS IN
CHEMISTRY

14. Synthesis of a dye stuff intermediate and an azo dye

In this experiment students are divided into two groups, each of which follows a different route for the synthesis of 4-bromoaminobenzene (*para*-bromoaniline), a key intermediate in the synthesis of an azo dye. Both groups then use their product to synthesise the azo dye. The yields and costs of each method are compared.

This experiment is used to help students improve their experimental planning skills. It should take each group four three-hour laboratory sessions to complete the work. Students are divided into two groups and given the experimental procedure in advance of the session. They are then required to hand in a schedule of how they plan to fit the work into the time available before they start work in the laboratory.

This experiment is suitable for:

■ second year students

■ four three-hour sessions

■ group work

Activity type

| formal | ○ | experimental | ○ | divergent | ● | investigatory | ○ |

Skills

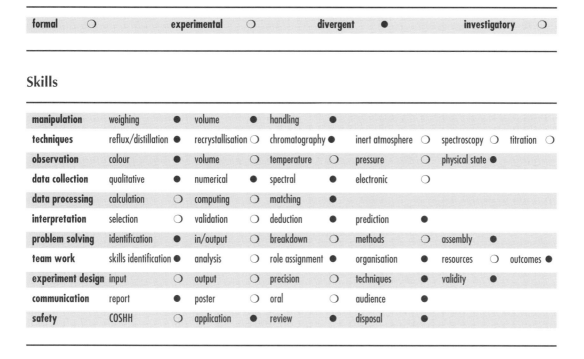

manipulation	weighing ●	volume ●	handling ●			
techniques	reflux/distillation ●	recrystallisation ○	chromatography ●	inert atmosphere ○	spectroscopy ○	titration ○
observation	colour ●	volume ○	temperature ○	pressure ○	physical state ●	
data collection	qualitative ●	numerical ●	spectral ●	electronic ○		
data processing	calculation ○	computing ○	matching ●			
interpretation	selection ○	validation ○	deduction ●	prediction ●		
problem solving	identification ●	in/output ○	breakdown ○	methods ○	assembly ●	
team work	skills identification ●	analysis ○	role assignment ●	organisation ●	resources ○	outcomes ●
experiment design	input ○	output ○	precision ○	techniques ●	validity ●	
communication	report ●	poster ○	oral ○	audience ●		
safety	COSHH ○	application ●	review ●	disposal ●		

PROGRESSIVE
DEVELOPMENT
OF PRACTICAL
SKILLS IN
CHEMISTRY
■

Synthesis of a dye stuff intermediate and an azo dye

In this activity, 4-bromoaminobenzene (*para*-bromoaniline) is synthesised by one of two routes and is then used to synthesise an azo dye. The yields and costs of each route are compared in order to determine which is the most useful.

The experiment takes four three-hour laboratory sessions to complete. A written plan indicating the way in which the laboratory session will be organised must be drawn up before starting.

Equipment

Route A
- 25 cm^3 measuring cylinder
- 250 cm^3 Quickfit round-bottomed flask
- 500 cm^3 Quickfit round-bottomed flask
- water bath
- 500 cm^3 beaker
- 500 cm^3 Buchner flask
- scissors or tin snips
- 250 cm^3 separating funnel
- TLC equipment

Route B
- Bunsen burner
- ceramic-centred wire gauze
- insulating mat
- tripod
- 50 cm^3 beaker

Routes A and B

- 10 cm^3 measuring cylinder
- 100 cm^3 measuring cylinder
- 100 cm^3 Quickfit round-bottomed flask
- Quickfit water condenser and tubing
- reduction adaptor
- anti-bumping granules
- heating mantle
- spatula
- 100 cm^3 beakers
- 250 cm^3 beakers
- glass rod
- Buchner funnel
- 100 cm^3 Buchner flask
- 250 cm^3 Buchner flask
- $3 \times 50 \text{ cm}^3$ conical flasks
- $4 \times 100 \text{ cm}^3$ conical flasks
- $3 \times 250 \text{ cm}^3$ conical flasks
- filter paper
- cotton wool
- glass filter funnel
- watch glass
- ice-water bath
- Hirsch funnel
- steam bath
- vacuum source
- thermometer (-10 °C to 110 °C)
- 100 cm^3 wide-mouthed conical flask
- dropping pipette
- melting point apparatus

Reagents

Route A

- concentrated sulfuric acid

- concentrated nitric acid

- bromobenzene

- granulated tin

- hydrochloric acid (2 mol dm^{-3})

- dichloromethane

- anhydrous magnesium sulfate

- sodium hydroxide pellets

- petroleum ether (boiling range 60–80 °C)

Route B

- aminobenzene

- glacial ethanoic acid

- ethanoic anhydride

- zinc dust

- bromine (in ethanoic acid solution)

- aqueous sodium hydrogen sulfite

- sodium hydroxide solution (5%)

- universal indicator paper

Routes A and B

- concentrated hydrochloric acid

- methylated spirit

- sodium nitrite

- 2-naphthol

Safety

Route A

Concentrated sulfuric acid: Burns skin and eyes. Causes severe damage if taken by mouth. Toxic by inhalation of fumes or mist.

Concentrated nitric acid: Corrosive. Burns skin and eyes. Violent reaction with organic compounds.

Bromobenzene: Irritating to skin and eyes.

Granulated tin: Can cause physical damage to eyes and skin.

Dichloromethane: Highly volatile and flammable. Will pressurise a closed vessel when shaken. Will degrease skin on contact. Harmful by inhalation. Irritating to eyes. Toxic if taken by mouth.

Anhydrous magnesium sulfate: Harmful by inhalation, in contact with skin and if swallowed.

Sodium hydroxide pellets: Corrosive. Rapidly absorb moisture from air forming a highly corrosive and toxic solution.

Petroleum ether: Highly flammable.

4-Bromonitrobenzene: Harmful. Irritant.

Route B
Aminobenzene: Toxic.

Ethanoic acid and ethanoic anhydride: Flammable. Corrosive. Irritating vapour.

Bromine in ethanoic acid solution: Toxic. Very corrosive. Vapour is a severe lung and eye irritant.

Sodium hydrogen sulfite solution: Irritant.

Sodium hydroxide solution (5%): Irritating to skin. Severely irritating to eyes.

4-Bromo-N-phenylacetamide: Toxic. Irritant.

Routes A and B
Methylated spirit: Highly flammable. Toxic.

Concentrated hydrochloric acid: Highly corrosive. Gives off choking fumes.

Sodium nitrite: Oxidising. Toxic.

2-Naphthol: Skin irritant. Harmful.

4-Bromoaminobenzene: Toxic. Irritant.

Azo dye product: Stains skin. Possible carcinogen.

Route A

Nitration of bromobenzene

1 Place 20 cm^3 of concentrated nitric acid in a 250 cm^3 Quickfit round-bottomed flask. Cool the flask in an ice-water bath, being careful not to let any water enter it. Add 20 cm^3 of concentrated sulfuric acid in approximately 5 cm^3 portions, swirling to mix the contents and cooling after each addition. Allow the contents of the flask to reach room temperature.

2 Attach a water condenser vertically to the round-bottomed flask, and support the flask and condenser with a retort stand and clamp in a fume cupboard. Do not run water through the condenser yet. Add 10 cm^3 (14.9 g) of bromobenzene in 2–3 cm^3 portions down the condenser over a period of 15 minutes, swirling after each addition. If the flask becomes too hot to touch, cool it briefly in an ice-water bath. Run water through the condenser, and heat the reaction mixture on a steam bath for 30 minutes.

3 Allow the reaction mixture to cool to room temperature, and then pour it, with vigorous stirring, into a mixture of ice (about 100 cm^3 – loosely packed) and water (200 cm^3) in a 500 cm^3 beaker. Transfer some of the water from the beaker to the reaction flask to rinse it out, and then pour this back into the beaker. When all the ice has melted, filter off the precipitated solid under suction. Wash with water (2 x 50 cm^3).

4 Recrystallise the solid product using methylated spirit. Allow to dry in air, or in an oven at no more than 80 °C. Record the percentage yield, melting point and infrared spectrum of the product.

Reduction of 4-bromonitrobenzene

The reduction procedure below is for 5 g of 4-bromonitrobenzene. The procedure may need to be scaled up or down, depending on how much of the compound has been synthesised.

RS•C

1   Crush the dry 4-bromonitrobenzene into a fine powder and weigh it. For each 5 g amount of solid starting material, weigh out 10 g of granulated tin. The tin must be cut into very small pieces using scissors or tin snips. In a fume cupboard, place the tin in a 250 cm^3 Quickfit round-bottomed flask and add concentrated hydrochloric acid (5 cm^3). Swirl the flask for 1–2 minutes and then decant the hydrochloric acid into a 250 cm^3 beaker half full of water.

Cover the tin with methylated spirit (20 cm^3) and water (80 cm^3), then add the powdered 4-bromonitrobenzene. Fit the flask with an air condenser. Add 2 cm^3 of concentrated hydrochloric acid through the condenser. Heat the flask on a steam bath for 20 minutes. Add a further 2 cm^3 of concentrated hydrochloric acid and continue heating for a further 60 minutes. Swirl the contents of the flask occasionally during the heating period.

2   Cool the reaction flask under a tap, allow the heavier part of the solid to settle out, and decant the remaining aqueous layer into a 250 cm^3 separating funnel. Wash the residual solid by swirling it with two portions of 2 mol dm^{-3} hydrochloric acid (2 x 25 cm^3), each time allowing the solid to settle and decanting the washings into a 500 cm^3 beaker. Wash the inside of the condenser with dichloromethane. Do this by slowly pouring 50 cm^3 of dichloromethane down the condenser into the flask. Swirl the flask, allow the solid to settle and decant the organic liquid into the same beaker. Transfer the contents of the beaker (both layers) to the separating funnel. Shake the stoppered funnel to ensure thorough mixing of the layers. The pressure in the funnel should be relieved from time to time. Allow the layers to separate and run out the lower (organic) layer into a 100 cm^3 conical flask. Keep the aqueous layer in the flask – it will be needed for the next part of the experiment. Dry the organic layer with anhydrous magnesium sulfate until there is no hint of cloudiness in the solution. Label and save this solution for part **6**.

3   Run the upper (aqueous) layer into a 500 cm^3 Quickfit round-bottomed flask, and add 200 cm^3 of 1 mol dm^{-3} aqueous sodium hydroxide (8 g of NaOH pellets in 200 cm^3 water) in 50 cm^3 portions, swirling after each addition. Check the pH of the mixture using indicator paper. If it is less than pH 14 add a few pellets of solid sodium hydroxide to increase the alkalinity. Fit the flask with a splash head and a long or double surface water condenser, and add a few anti-bumping granules to the liquid. Distil over a Bunsen burner. Collect the distillate until it is clear, then collect a further 30-50 cm^3. At no time should the volume of the liquid in the round-bottomed flask fall below 100 cm^3. If the product begins to crystallise in the condenser turn off the water flow until the crystals start to melt.

4   Extract the steam distillate using two separate portions of dichloromethane (2 x 30 cm^3). Collect the organic fractions in a 100 cm^3 conical flask. If an emulsion forms, dissolve a little solid sodium chloride in the aqueous layer and shake the funnel gently. Dry the organic layer with anhydrous magnesium sulfate and filter the dried solution through a fluted filter paper into a pre-weighed 250 cm^3 Quickfit round-bottomed flask. Evaporate the solvent using a rotary evaporator. Reweigh the flask to obtain a crude yield.

5 There are two ways to carry out this recrystallisation, depending on the condition of your sample.

A – for a damp or oily sample
Add methylated spirit (6 cm^3 for every 2 g of crude product) to the round-bottomed flask containing the crude product, and warm the flask on a steam bath until all the solid has dissolved. Pour the hot solution into a 50 cm^3 conical flask, and add an equal volume of water. Warm the flask again on the steam bath. If the solution remains cloudy when swirled, add drops of methylated spirit until it clears; if the solution is completely clear, add drops of water until it becomes cloudy.

B – for a dry sample
Fit the round-bottomed flask with a water-cooled condenser. Add 20 cm^3 of petroleum ether (boiling range 60–80 °C) carefully down the condenser, and reflux vigorously over a water bath for two minutes. If some solid remains undissolved add another 5 cm^3 of petroleum ether and reflux again. Decant the hot solution into a 100 cm^3 conical flask, leaving any insoluble material behind.

In both cases allow the flask to cool to room temperature, and then cool it in an ice-water bath for at least 10 minutes after the first crystals form. Filter off the crystals under vacuum. Record the melting point, yield and infrared spectrum of the product.

6 Filter the dichloromethane extract from part **2** through a fluted filter paper. Evaporate the solvent on a rotary evaporator. The residual material should be unreacted starting material, 4-bromonitrobenzene. Confirm this by either thin layer chromatographic analysis or infrared spectroscopy. Record the yield of unreacted starting material. Recalculate the percentage yield of 4-bromoaminobenzene after subtracting the amount of 4-bromonitrobenzene recovered from the amount of starting material used for the reaction.

Route B

Acetylation of aminobenzene

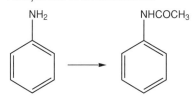

1 Place 5 cm^3 (5.1 g) of aminobenzene, 5 cm^3 (5.4 g) of ethanoic anhydride and 5 cm^3 (5.2 g) of ethanoic acid successively into a 100 cm^3 Quickfit round-bottomed flask fitted for reflux with a water-cooled condenser. Add a few anti-bumping granules and about 0.05 g of zinc dust, and reflux the mixture gently for 30 minutes on a heating mantle.

2 Pour the hot liquid in a thin stream into a 250 cm^3 beaker containing 100 cm^3 of water, cooled by adding a little ice, stirring continually during the addition. Cool the beaker in an ice-water bath to precipitate the crude solid, which should be filtered under suction. Wash the

crystals with cold water, drain, and then transfer to a 250 cm^3 conical flask.

3   Recrystallise the product from a mixture of water and methylated spirit (about 4 cm^3 of methylated spirit and 200 cm^3 of water). Place the conical flask on a ceramic-centred gauze resting on top of a tripod. Add 25 cm^3 of solvent mixture at a time, swirling and heating the flask until the solid dissolves. If an oil forms, make sure all the globules have dissolved, and that the solution is homogeneous. If a lot of zinc dust remains, decant the hot solution rapidly into a second conical flask through a small piece of cotton wool in a glass filter funnel. Once crystals have formed, cool the flask for a further 10 minutes in an ice-water bath. Filter off the crystals under suction, and rinse with a little cold water. Dry in an oven at not more than 80 °C for an hour.

Calculate the yield, measure the melting point and record the infrared spectrum of the product.

Bromination of N-phenylacetamide

This reaction should be carried out in a fume cupboard, and gloves should be worn whilst handling bromine solutions.

1   In a 100 cm^3 conical flask dissolve 4.5 g of N-phenylacetamide in 20 cm^3 of glacial ethanoic acid. Slowly add 7 cm^3 of bromine in ethanoic acid to the solution using a dropping pipette. Stir the solution continually during addition. The flask may need to be cooled in a water bath. Allow the mixture to stand at room temperature for at least 30 minutes with occasional shaking.

2   If the reaction mixture is still appreciably coloured by residual bromine, add just sufficient aqueous sodium hydrogen sulfite solution to remove the orange colour on swirling. Pour into 100 cm^3 of water in a 250 cm^3 conical flask, and rinse out the original flask with some water. Stir the mixture well, making sure that all the lumps of precipitate are crushed and dispersed. The flask can now be removed from the fume cupboard. Filter the precipitate under suction and wash with cold water.

3   Transfer the solid to a 100 cm^3 conical flask, and recrystallise from the minimum volume of 2:1 methylated spirit/water, using a steam bath. Calculate the yield, measure the melting point and record the infrared spectrum of the product.

Hydrolysis of 4-bromo-N-phenylacetamide

1   Place around 3.5– 4 g of 4-bromo-N-phenylacetamide (accurately weighed) in a 100 cm^3 Quickfit round-bottomed flask, and add a 1:1 mixture of water (15 cm^3) and methylated spirit (15 cm^3). Fit the flask with a water-cooled condenser, add a few anti-bumping granules, and heat the contents to boiling on a heating mantle. Using a dropping pipette, add 10 cm^3 of concentrated hydrochloric acid in small portions down the condenser, at such a rate that boiling is continuous throughout the addition. Reflux for 40–60 minutes.

2   Cool the flask under the tap, and then decant the solution into a second 100 cm^3 Quickfit round-bottomed flask, being careful to leave the anti-bumping granules behind. Evaporate the solvent on a rotary evaporator until the reaction mixture is reduced to about half its original volume. Pour the residual reaction mixture into 25 cm^3 of ice-water in a 250 cm^3 conical flask, and rinse out the flask with a little of the ice-water mixture. Add 5% sodium hydroxide solution, with vigorous stirring, until the mixture becomes persistently cloudy. Test that it is just alkaline using indicator paper. The 4-bromoaminobenzene should separate out as fine crystals. If it comes down as an oil, add more ice and disperse the globules in the cold solution, where they should crystallise.

   When all the ice has melted, filter the solid product under suction, and dry as far as possible in the funnel. Record the crude yield.

3   Recrystallise with methylated spirit, using a steam bath to warm the solution. When all the solid has dissolved add an equal volume of water and warm again on the steam bath. If the solution remains cloudy when swirled, add drops of methylated spirit until it clears; if the solution is already clear add drops of water until it becomes cloudy. Allow the solution to cool, then cool further in an ice-water bath. Filter off the crystals under suction. Calculate the yield, measure the melting point and record the infrared spectrum of the product.

Routes A and B

Diazotization and coupling of 4-bromoaminobenzene

This procedure is based on 0.86 g of 4-bromoaminobenzene. It may need to be scaled up or down, depending on the quantity of product available.

1 Dissolve 4-bromoaminobenzene (0.86 g) in a mixture of concentrated hydrochloric acid (4 cm^3) and water (10 cm^3) in a 50 cm^3 beaker. Warm the mixture to dissolve the amine, then cool in an ice-water bath with constant stirring. Continue to cool until the temperature of the contents falls below 5 °C. Crystals of 4-bromoaminobenzidinium chloride will precipitate in a slush.

Dissolve sodium nitrite (0.4 g) in water (4 cm^3) in a 25 cm^3 conical flask, and cool in the ice-water bath.

Dissolve 2-naphthol (0.72 g) in 5% aqueous sodium hydroxide solution (10 cm^3) in a wide mouthed 100 cm^3 conical flask, and cool in the ice-water bath.

2 Slowly add the cold sodium nitrite solution with a dropping pipette to the cold slush of 4-bromoaminobenzidinium chloride, swirling after each addition. Keep the temperature below 5 °C, adding small pieces of ice to the reaction mixture if necessary. Allow this solution of the diazonium salt to stand for at least 5 minutes in the ice-water bath, swirling occasionally.

Slowly add the diazonium salt solution dropwise to the 2-naphthol solution, again maintaining the temperature below 5 °C. Allow the reaction mixture to stand in the ice-water bath for a further 15 minutes with occasional swirling.

3 Filter the suspension through a Buchner funnel under gentle suction. Wash the crystals with water (20 cm^3) then methylated spirit (5 cm^3). Drain the crystals well, then allow to dry in air. Calculate the yield, measure the melting point and record an infrared spectrum of the product.

Results

Assemble the results into a table, giving the yield in grams, the percentage molar yield for each stage of the reaction, and the melting point of the product. Note the melting points recorded in the literature and calculate the overall yield of 4-bromoaminobenzene. As both routes A and B are linear syntheses, the overall fractional yield for each route is obtained by multiplying together the fractional yields for the individual stages. Assemble a second table with the results from someone who followed the alternative route.

Calculate the cost of each route, using the prices of reagents listed in the table below. In order to do this, the cost of the intermediates isolated in this experiment will need to be calculated.

Reagent	Unit cost (May 1998 prices)
concentrated HNO_3	£13.70 per 2.5 dm^{-3}
concentrated H_2SO_4	£11.10 per 2.5 dm^{-3}
bromobenzene	£10.20 per 0.5 dm^{-3}
methylated spirit	£11.40 per 2.5 dm^{-3}
tin	£11.40 per 0.1 kg
concentrated HCl	£14.90 per 4.0 dm^{-3}
dichloromethane	£17.80 per 2.5 dm^{-3}
magnesium sulfate	£4.30 per 0.5 kg
sodium hydroxide	£10.70 per 0.5 kg
aminobenzene	£9.30 per 1.0 dm^{-3}
ethanoic anhydride	£10.80 per 2.5 dm^{-3}
ethanoic acid	£12.20 per 2.5 dm^{-3}
zinc dust	£9.60 per 1.0 kg
bromine	£24.40 per 0.5 dm^{-3}

The commercial cost of 4-bromoaminobenzene is £15.57 for 0.1 kg. How does this compare with the calculated value? If there is any difference, why might this be?

Bearing in mind the costs and overall percentage yields, which route would be better to use for the synthesis of 4-bromoaminobenzene?

D14

Synthesis of a dye stuff intermediate and an azo dye

In this activity 4-bromoaminobenzene (*para*-bromoaniline) is synthesised by two routes and is used to synthesise an azo dye. The yields and costs of each route are compared in order to determine which is the most useful.

The experiment takes four three-hour laboratory sessions to complete. Before starting, students must complete a written plan showing the way in which the laboratory session will be organised. The pre-laboratory part of this experiment is important, as students have been given little theoretical background about the reaction in their notes.

The structure of the azo dye synthesised in this experiment is:

The N=N bridge can be synthesised by reaction of a diazonium salt and an aromatic species via an electrophilic substitution reaction. In this case 2-naphthol and 4-bromoaminobenzene will react to form the dye.

2-naphthol is readily available and therefore only 4-bromoaminobenzene must be synthesised. Two approaches are possible starting with either bromobenzene (route A) or with aminobenzene (route B).

In Route A, the amine is made by first synthesising the corresponding nitro compound, and reducing with tin and HCl to bromonitrobenzene. Although some 2-bromonitrobenzene, and a little 3-bromonitrobenzene are formed, the 4-bromonitrobenzene product is easily isolated.

In Route B, the amine group is converted to an amide, and brominated to form a reasonable yield of the 4-bromo-N-phenylacetamide product. The amide is then hydrolysed to reform the amine.

Students should be asked which factors they think are important in choosing a reaction route. Answers should include:

■ maximum yield

■ time

■ ease of isolating compound

■ cost of reagents and solvents

■ hazards associated

In a multi-step synthesis, yields fall very quickly at each step. Calculating the overall yield for a three step reaction if the yield at each step is 70% might be useful.

Organisation of work

	Route A	**Route B**
Session 1	Nitration of bromobenzene; isolation and recrystallisation of 4-bromonitrobenzene.	Preparation, isolation and purification of N-phenylacetamide; bromination of N-phenylacetamide.
Session 2	Reduction of 4-bromonitrobenzene; IR and melting point (mp) of 4-bromonitrobenzene.	Isolation and purification of 4-bromo-N-phenylacetamide; IR and mp of 4-bromo-N-phenylacetamide.
Session 3	Isolation and steam distillation of 4-bromoaminobenzene; extraction and recrystallisation of 4-bromoaminobenzene.	Hydrolysis of 4-bromo-N-phenyl-acetamide; isolation and recrystallisation of 4-bromoamino-benzene.
Session 4	IR and mp of 4-bromoaminobenzene; preparation and isolation of azo dye.	IR and mp of 4-bromoaminobenzene; preparation and isolation of azo dye.

Route A

The acids and bromobenzene must be thoroughly swirled between each addition, otherwise a build-up of unreacted material can lead to a sudden, very rapid reaction. The procedure should be carried out in a fume cupboard to prevent the escape of NO_2 into the laboratory.

The product must be washed thoroughly to remove acid on the outside of the crystals, otherwise the recrystallisation solvent becomes acidic and oxidising.

If the reduction of 4-bromonitrobenzene shows no signs of proceeding, the addition of more finely divided tin should solve the problem.

Route B

This is the easier, safer, but longer route.

Routes A and B

Students should have prepared about 1 g of 4-bromoaminobenzene with which to carry out this reaction, but starting quantities down to about 0.4 g give good results.

Melting points of products

The melting points of the products of routes A and B are given below:

- 4-bromonitrobenzene   125–127 °C
- N-phenylacetamide   113–115 °C
- 4-bromo-N-phenylacetamide   167–169 °C
- 4-bromoaminobenzene   62–64 °C

Ideas for further investigation

Thin layer chromatographic investigation of the products of nitration of bromobenzene.

Preparation of 2,4-dinitrobromobenzene.

Direct bromination of aminobenzene.

T14

Synthesis of a dye stuff intermediate and an azo dye

Equipment

Route A
- 25 cm^3 measuring cylinder
- 250 cm^3 Quickfit round-bottomed flask
- 500 cm^3 Quickfit round-bottomed flask
- water bath
- 500 cm^3 beaker
- 500 cm^3 Buchner flask
- scissors or tin snips
- 250 cm^3 separating funnel
- TLC equipment

Route B
- Bunsen burner
- ceramic-centred wire gauze
- insulating mat
- tripod
- 50 cm^3 beaker

Routes A and B
- 10 cm^3 measuring cylinder
- 100 cm^3 measuring cylinder
- 100 cm^3 Quickfit round-bottomed flask
- Quickfit water condenser and tubing
- reduction adapter
- anti-bumping granules
- heating mantle
- spatula
- 100 cm^3 beakers
- 250 cm^3 beakers

- glass rod
- Buchner funnel
- 100 cm^3 Buchner flask
- 250 cm^3 Buchner flask
- 3 x 50 cm^3 conical flasks
- 4 x 100 cm^3 conical flasks
- 3 x 250 cm^3 conical flasks
- filter paper
- cotton wool
- glass filter funnel
- watch glass
- ice-water bath
- Hirsch funnel
- steam bath
- vacuum source
- thermometer (-10 °C to 110 °C)
- 100 cm^3 wide mouthed conical flask
- dropping pipette
- melting point apparatus

Reagents

Route A
- concentrated sulfuric acid
- concentrated nitric acid
- bromobenzene
- granulated tin
- hydrochloric acid (2 mol dm^{-3})
- dichloromethane
- anhydrous magnesium sulfate
- sodium hydroxide pellets
- petroleum ether (boiling range 60–80 °C)

Route B

■ aminobenzene

■ glacial ethanoic acid

■ ethanoic anhydride

■ zinc dust

■ bromine (in ethanoic acid solution)

■ aqueous sodium hydrogen sulfite

■ sodium hydroxide solution (5%)

■ universal indicator paper

Routes A and B

■ concentrated hydrochloric acid

■ methylated spirit

■ sodium nitrite

■ 2-naphthol

15. Synthesis of an epoxide

This is a 'puzzle' type of organic chemistry experiment. The experiment investigates the stereochemistry of two reactions: the conversion of an alkene to a bromohydrin by reaction with N-bromosuccinimide, and the subsequent reaction of the bromohydrin with acid to form an epoxide. The stereochemistry of each product from each part of the reaction is determined by the melting point. The activity can be completed in approximately six hours.

This experiment is suitable for:

■ first year or early second year

■ approximately six hours

■ group or individual work

Activity type

formal	○	experimental	●	divergent	○	investigatory	○

Skills

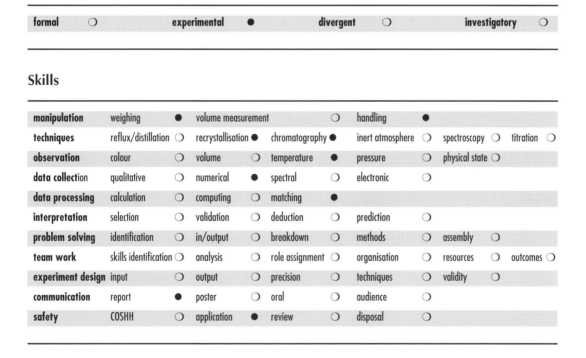

manipulation	weighing	●	volume measurement		○	handling	●				
techniques	reflux/distillation	○	recrystallisation ●	chromatography ●		inert atmosphere	○	spectroscopy	○	titration	○
observation	colour	○	volume	○	temperature ●	pressure	○	physical state ○			
data collection	qualitative	○	numerical	●	spectral	○	electronic	○			
data processing	calculation	○	computing	○	matching	●					
interpretation	selection	○	validation	○	deduction	○	prediction	○			
problem solving	identification	○	in/output	○	breakdown	○	methods	○	assembly	○	
team work	skills identification ○		analysis	○	role assignment ○	organisation	○	resources	○	outcomes ○	
experiment design	input	○	output	○	precision	○	techniques	○	validity	○	
communication	report	●	poster	○	oral	○	audience	○			
safety	COSHH	○	application	●	review	○	disposal	○			

Synthesis of an epoxide

Epoxides are a synthetically useful class of compounds. There are a number of naturally occurring chiral epoxides. In this experiment a two-step synthesis of an epoxide is carried out, starting with halohydrin formation from an alkene, followed by intramolecular Williamson ether synthesis to form an epoxide.

This experiment explores whether each of the reactions undertaken is stereospecific and, if so, which stereoisomer is formed.

Treatment of an alkene with aqueous N-bromosuccinimide (NBS) affords a bromohydrin.

The stereoselectivity of this reaction is investigated by treating *trans*-1,2-diphenylethene (*trans*-stilbene) with NBS and H_2O in dimethyl sulfoxide (DMSO). This gives two possible diastereomers, whose melting points differ by 30 °C, so the products can readily be differentiated by a melting point determination.

If the bromohydrin is treated with a base, HBr is eliminated to form an epoxide.

For this experiment, the bromohydrin is reacted with potassium carbonate in methanol, on a microscale level. This reaction can proceed with either retention or inversion of configuration, to produce one of two possible diastereomers. The melting points of the products differ by 25 °C and can therefore be used to distinguish the products.

Equipment

- spatula
- weighing bottle with lid
- pipettes to measure 0.5 cm^3 and 15 cm^3
- pipette fillers
- 50 cm^3 conical flask
- steam bath
- stirring rod
- 4 x 100 cm^3 beakers
- 2 x 250 cm^3 beakers
- 500 cm^3 beaker
- 250 cm^3 separating funnel
- 50 cm^3 measuring cylinder
- Buchner flask
- Buchner funnel
- funnel
- 250 cm^3 round-bottomed flask
- 2 x test-tubes (20 x 150 mm)
- corks or rubber stoppers
- thin layer chromatography plates
- watch glass (large enough to fit over beaker)
- spotters or capillary tubes
- Pasteur pipettes
- melting point apparatus
- magnetic stirrer
- rotary evaporator
- ultra-violet light or heat gun/hairdryer
- analytical balance

Reagents

- *trans*-1,2-diphenylethene
- dimethyl sulfoxide
- N-bromosuccinimide
- diethyl ether
- celite

- saturated brine

- anhydrous magnesium sulfate

- petroleum ether (boiling range 60–80 °C)

- anhydrous potassium carbonate

- methanol

- ethyl ethanoate

- phosphomolybdic acid (20% in ethanol)

Safety

Dimethyl sulfoxide: Irritant to skin. Can damage eyes. Harmful if absorbed through skin.

N-bromosuccinimide: Irritant to skin, eyes and respiratory system.

Diethyl ether: Extremely flammable. May be harmful by ingestion and inhalation. Irritating to eyes and degreases skin.

Celite: Harmful if inhaled as dust.

Anhydrous magnesium sulfate: Harmful by inhalation, in contact with skin and if swallowed.

Petroleum ether: Extremely flammable. Toxic by inhalation and if swallowed. Irritating to eyes, respiratory system and skin.

Anhydrous potassium carbonate: Harmful by inhalation, in contact with skin and if swallowed. Irritating to eyes, respiratory system and skin.

Methanol: Highly flammable. Toxic by inhalation and if swallowed. Can cause delayed damage to eyes if ingested.

Ethyl ethanoate: Highly flammable. Prolonged inhalation can cause kidney and liver damage. Liquid and vapour irritate eyes.

Phosphomolybdic acid in ethanol (20%): Contact with combustible material may cause fire. Toxic by inhalation, in contact with skin and if swallowed. Causes burns. Highly flammable.

trans-1,2-Diphenylethene: Harmful by ingestion. Irritating to eyes.

trans-1,2-Diphenylethene oxide: May be harmful by ingestion, inhalation and skin absorption. May cause eye and skin irritation.

Procedure

Synthesis of bromohydrin

1   Place 1.0 g of *trans*-1,2-diphenylethene in a 50 cm^3 conical flask, followed by 0.5 cm^3 of water, and 15 cm^3 of DMSO. To dissolve all of the alkene, heat the solution over a steam bath or add more DMSO, if necessary.

2   Add 2 molar equivalents of NBS (relative to the alkene) in small portions over several minutes.

3   Stir the resulting solution for 30 minutes, then pour into a large beaker containing 50 cm^3 of ice-cold water.

4   Transfer the resulting slurry to a separating funnel and add 20 cm^3 of water. Extract using two 20 cm^3 aliquots of diethyl ether.

5   Remove any undissolved solid by vacuum filtering the contents of the separating funnel through a thin layer of celite. Additional diethyl ether might be needed, as evaporation of the diethyl ether may cause the product to precipitate.

6   Separate the layers. Extract the aqueous layer with two 15 cm^3 aliquots of diethyl ether.

7   Combine the ether extracts, and wash with 50 cm^3 of water, followed by 50 cm^3 of saturated brine. Dry over anhydrous $MgSO_4$. Remove the $MgSO_4$ by gravity filtration into a round-bottomed flask.

8   Remove the diethyl ether by rotary evaporation. Record the yield of crude product.

Recrystallisation of bromohydrin

Recrystallise using a minimum amount of petroleum ether (bp 60–80 °C). A small amount of solid will remain undissolved, hot filter the solution to remove it.

Synthesis of epoxide

1   Place 150 mg of dry bromohydrin into a 20 x 150 mm test-tube, followed by 125 mg of anhydrous potassium carbonate and 2 cm^3 of methanol. Cork the tube.

2   Shake the tube at frequent intervals.

3   Carry out a thin layer chromatographic analysis of the reaction mixture every five minutes, using an elution mixture of 20:1 petroleum ether:ethyl ethanoate. Both the product and the reactant spots will be visible under ultra-violet light. They can also be visualised by dipping the plate in the phosphomolybdic acid solution, and then gently heating it with a heat gun.

4   When the reaction is complete, add 3 cm^3 of water, followed by 7 cm^3 of petroleum ether. Re-cork the tube, and shake.

5   Use a Pasteur pipette to transfer the top, organic layer into a second test-tube, and treat with anhydrous $MgSO_4$. Remove the drying agent by gravity filtration, then remove the solvent on a rotary evaporator. Record the yield.

Melting point (mp) determination

There are two possible bromohydrin products: (±)-*erythro*-2-bromo-1,2-diphenylethanol (mp 83–84 °C), and (±)-*threo*-2-bromo-1,2-diphenylethanol (mp 51–52 °C).

There are also two possible epoxide products: (±)-*trans*-1,2-diphenylethene oxide (mp 65–67 °C), and (±)-*cis*-1,2-diphenylethene oxide (mp 38–40 °C).

Synthesis of an epoxide

This experiment explores whether each of the reactions undertaken is stereospecific and, if so, which stereoisomer is produced. The possible stereoisomeric products for each reaction can be easily differentiated by a melting point determination.

Synthesis of bromohydrin
The product is (±)-*erythro*-2-bromo-1,2-diphenylethanol.

This part of the experiment should take approximately four hours (excluding time for recrystallisation). The synthesis itself should take approximately three hours for students to carry out. Recrystallisation of the product and the next part of the experiment can be carried out in another session if necessary.

Synthesis of epoxide
The product is (±)-*trans*-1,2-diphenylethene oxide.

This part of the experiment should take approximately two hours.

The stereochemistry of the epoxide can also be investigated using NMR. (^{1}H and ^{13}C).

PROGRESSIVE
DEVELOPMENT
OF PRACTICAL
SKILLS IN
CHEMISTRY

The mechanism for the reactions carried out by the students is as follows:

Reference

This experiment was adapted from J Ciaccio, *J. Chem. Educ.*, 1995, **72**, 1037.

T15

Synthesis of an epoxide

Equipment

- spatula
- weighing bottle with lid
- pipettes to measure 0.5 cm^3 and 15 cm^3
- pipette fillers
- 50 cm^3 conical flask
- steam bath
- stirring rod
- 4 x 100 cm^3 beakers
- 2 x 250 cm^3 beakers
- 500 cm^3 beaker
- 250 cm^3 separating funnel
- 50 cm^3 measuring cylinder
- Buchner flask
- Buchner funnel
- funnel
- 250 cm^3 round-bottomed flask
- 2 x test-tubes (20 x 150 mm)
- corks or rubber stoppers
- thin layer chromatography plates
- watch glass (large enough to fit over beaker)
- spotters or capillary tubes
- Pasteur pipettes
- melting point apparatus
- magnetic stirrer
- rotary evaporator
- ultra-violet light or heat gun/hairdryer
- analytical balance

Reagents

- *trans*-1,2-diphenylethene
- dimethyl sulfoxide (DMSO)
- N-bromosuccinimide (NBS)
- diethyl ether
- celite
- saturated brine
- anhydrous magnesium sulfate
- petroleum ether (boiling range 60–80 °C)
- anhydrous potassium carbonate
- methanol
- ethyl ethanoate
- phosphomolybdic acid (20% in ethanol)

RS•C
PROGRESSIVE
DEVELOPMENT
OF PRACTICAL
SKILLS IN
CHEMISTRY

119

16. Nickel catalyst

In this experiment students synthesise tetrakis(triethyl phosphite)nickel, Ni[P(OEt)$_3$]$_4$, and then investigate its use as an alkene isomerisation catalyst, using gas chromatographic analysis to follow the course of the reaction.
It includes a reaction that must be carried out under an inert atmosphere.

This experiment is suitable for:

■ second year students

■ approximately five hours

■ individuals/groups

Activity type

formal	○	experimental	●	divergent	○	investigatory	○

Skills

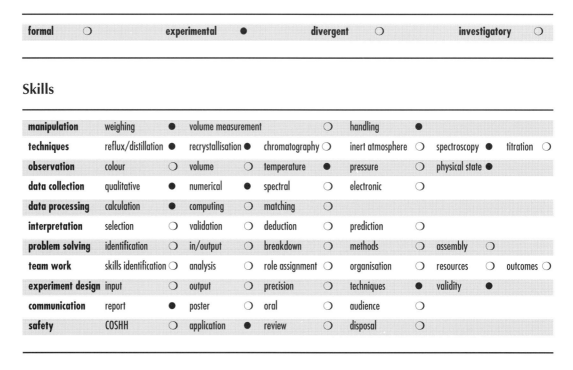

manipulation	weighing	●	volume measurement	○	handling	●				
techniques	reflux/distillation	●	recrystallisation ●	chromatography ○	inert atmosphere ○	spectroscopy ●	titration ○
observation	colour	○	volume	○	temperature	●	pressure	○	physical state ●
data collection	qualitative	●	numerical	●	spectral	○	electronic	○
data processing	calculation	●	computing	○	matching	○
interpretation	selection	○	validation	○	deduction	○	prediction	○
problem solving	identification	○	in/output	○	breakdown	○	methods	○	assembly	○
team work	skills identification ○	analysis	○	role assignment	○	organisation	○	resources	○	outcomes ○
experiment design	input	○	output	○	precision	○	techniques	●	validity	●
communication	report	●	poster	○	oral	○	audience	○
safety	COSHH	○	application	●	review	○	disposal	○

S16

Nickel Catalyst

In this activity a heterogeneous catalyst is prepared and its efficiency tested.

Equipment

- 100 cm^3 beaker
- 50 cm^3 measuring cylinder
- weighing bottles and lids
- spatula
- ice bath
- 10 cm^3 graduated pipette
- 20 cm^3 pipette
- 1 cm^3 pipette
- pipette filler
- syringe
- 100 cm^3 three-necked flask
- Buchner flask and funnel
- filter paper
- nitrogen inlet
- septum
- oil bubbler
- test-tubes
- micropipette
- stirrer bar
- magnetic stirrer
- vacuum desiccator
- gas chromatograph
- stop-clock
- melting point apparatus

Reagents

- nickel(II) chloride ($NiCl_2.6H_2O$)
- methanol
- triethylphosphite
- diethylamine
- nitrogen supply
- diethyl ether
- sulfuric acid (0.1 mol dm^{-3})
- 1-heptene

Safety

Nickel(II) chloride: Prolonged contact with skin can cause irritation.

Methanol: Highly flammable. Toxic by inhalation and if swallowed. Can cause delayed damage to eyes if ingested.

Triethylphosphite: Flammable. Harmful if swallowed. Irritating to skin, eyes and respiratory system.

Diethylamine: Highly flammable. Irritating to eyes and respiratory system. Do not empty into drains.

Diethyl ether: Toxic by inhalation. Extremely flammable.

Sulfuric acid (0.1 mol dm^{-3}): Irritating to skin and eyes. Causes burns.

1-Heptene: Highly flammable. Harmful by inhalation and if swallowed. Irritating to eyes, skin and respiratory system.

This experiment should be carried out in a fume cupboard.

Procedure

Synthesis of tetrakis(triethylphosphite)nickel, Ni[P(OEt)$_3$]$_4$

Place 2.6 g of $NiCl_2.6H_2O$ and 50 cm^3 of methanol into a 100 cm^3 beaker. Stir for five minutes, until all the nickel chloride has dissolved. Place in an ice bath, and add 9.4 cm^3 (9.1 g) of triethylphosphite over a two minute period. After stirring for five minutes, add 1.7 cm^3 of diethylamine. Add the diethylamine slowly dropwise, until the deep-red solution just begins to fade. The addition of too much diethylamine will result in a green nickel(II) contaminant. Stir the solution for ten minutes, keeping it in the ice bath. Remove the solid by filtering the solution under vacuum, and washing with ice-cold methanol (3 x 10 cm^3). Dry the solid under vacuum in a vacuum desiccator.

Determine the melting point of the product. This compound should be stored under vacuum or under nitrogen and away from light.

Catalytic isomerisation of 1-heptene

Fit a 100 cm^3 three-necked flask with a nitrogen inlet, septum, stirrer bar and oil bubbler. Place 71 mg of Ni[P(OEt)$_3$]$_4$ into the flask, and flush with nitrogen for several minutes. Add 20 cm^3 of diethyl ether via pipette, and bubble nitrogen through the solution for several minutes. Add 0.2 cm^3 of 1-heptene via syringe. Take a gas chromatogram of the solution – this will represent time zero. Cool the reaction mixture in an ice-water bath, and add 1.0 cm^3 of ice-cold H_2SO_4 in methanol via pipette. The addition of H_2SO_4 initiates the isomerisation reaction. Decrease the nitrogen flow to prevent the evaporation of heptene. The initial concentrations of the reactants are:

$[Ni[P(OEt)_3]_4]$: 4.6×10^{-3} mol dm^{-3}; $[H_2SO_4]$: 4.7×10^{-3} mol dm^{-3}; [1-heptene]: 0.067 mol dm^{-3}

Take 1 cm^3 aliquots of the reaction mixture via syringe at 4.5, 10, 17 and 29 minutes. Syringe the samples into test-tubes, and shake immediately in the air. This will quench the reaction by poisoning the nickel catalyst with oxygen. Inject 1 μl of each sample into the gas chromatograph. Use the data to plot a graph of mole percent heptene against time.

Nickel catalyst

In this activity a heterogeneous catalyst is prepared and its efficiency tested.

At the end of this experiment students should be able to:

- carry out a reaction under nitrogen; and

- use gas chromatography to follow the course of a reaction.

Synthesis of tetrakis(triethylphosphite)nickel
This product is a white solid, with a melting point of 108–109 °C.

Catalytic isomerisation of 1-heptene
Demonstration of the safe operation of a nitrogen cylinder may be necessary. At nickel concentrations of 0.003 mol dm^{-3} or less, care must be taken to exclude oxygen, as the catalyst is readily poisoned by oxygen.

The equilibrium product distribution of 1-heptene:2-heptene:3-heptene is 1:20:78.

Gas chromatography

Parameters for gas chromatography:

- Detector – flame ionisation detector

- Isothermal oven – 40 °C

- Capillary column – cross-linked methylsilicone, HP-1,
  12 m x 0.2 mm x 0.33 μm

Cis and *trans* isomers cannot be separated under these conditions.

Further investigation

Determine the effect of changing ligands on the isomerisation rate.

Determine the rate of decomposition of the nickel hydride.

The idea for this experiment came from K R Bradwhistell and J Lanza, *J. Chem. Educ.*, 1997, **74**, 579-581.

T16

PROGRESSIVE
DEVELOPMENT
OF PRACTICAL
SKILLS IN
CHEMISTRY
■

Nickel catalyst

Equipment

- 100 cm^3 beaker
- 50 cm^3 measuring cylinder
- weighing bottles and lids
- spatula
- ice bath
- 10 cm^3 graduated pipette
- 20 cm^3 pipette
- 1 cm^3 pipette
- pipette filler
- syringe
- 100 cm^3 three-necked flask
- Buchner flask and funnel
- filter paper
- nitrogen inlet
- septum
- oil bubbler
- test-tubes
- micropipette
- stirrer bar
- magnetic stirrer
- vacuum desiccator
- gas chromatograph
- stop-clock
- melting point apparatus

Reagents

- ■ nickel(II) chloride ($NiCl_2.6H_2O$)
- ■ methanol
- ■ triethylphosphite
- ■ diethylamine
- ■ nitrogen supply
- ■ diethyl ether
- ■ sulfuric acid (0.1 mol dm^{-3})
- ■ 1-heptene

17. Polyhalides

This activity develops a wide range of experimental and analytical methods and, in particular, develops skills in solving a synthetic problem. The main steps are the preparation of the interhalogen compound, ICl, and its subsequent reaction with a Group 1 metal bromide. In principle it is possible to produce any polyhalide anion, [XYZ⁻]. Because of the hazards involved with handling chlorine gas and ICl, along with the exacting nature of the analytical methods, the instructions for this experiment are very explicit in parts. A flow chart has been provided at the beginning of the experiment, to enable students to follow their work.

This experiment is most suitable for second year students, who should have enough experience in the laboratory to carry out the work safely. However, it can also be undertaken by small, well supervised groups of first year students. The work is best done by individuals, but can be tackled by pairs or groups. The experiment can be completed in three to four hours.

This experiment is suitable for:

■ second year students

■ three to four hours

■ individuals

Activity type

| formal | ○ | experimental | ● | divergent | ○ | investigatory | ○ |

Skills

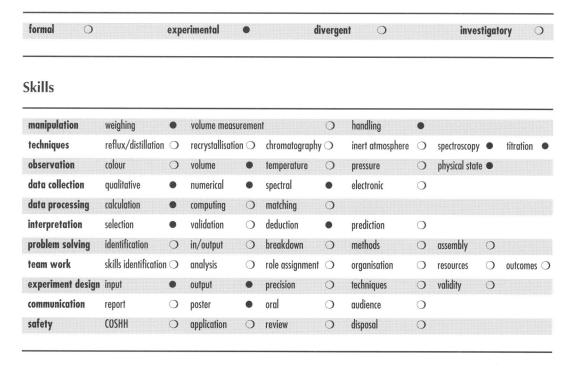

manipulation	weighing	●	volume measurement	○	handling	●				
techniques	reflux/distillation	○	recrystallisation ○	chromatography ○	inert atmosphere	○	spectroscopy ●	titration ●
observation	colour	○	volume	●	temperature	○	pressure	○	physical state ●
data collection	qualitative	●	numerical	●	spectral	●	electronic	○	
data processing	calculation	●	computing	○	matching	○			
interpretation	selection	●	validation	○	deduction	●	prediction	○	
problem solving	identification	○	in/output	○	breakdown	○	methods	○	assembly ○
team work	skills identification ○	analysis	○	role assignment ○	organisation	○	resources ○	outcomes ○
experiment design	input	●	output	●	precision	○	techniques	○	validity ○
communication	report	○	poster	●	oral	○	audience	○	
safety	COSHH	○	application	○	review	○	disposal	○	

RS•C

JOHN MOORES UNIVERSITY
AVRIL ROBARTS LRC
TITHEBARN STREET
LIVERPOOL L2 2ER
TEL. 0151 231 4022

Polyhalides

This activity has three main stages. Firstly, iodine monochloride, ICl, is prepared by reacting iodine with chlorine. This compound is then treated with caesium bromide to form a polyhalide salt. Finally, the polyhalide is identified and analysed.

Equipment

- internal-seal trap (dry)
- beaker to support trap
- glass funnel (dry)
- pestle and mortar (dry)
- weighing bottle (dry)
- 25 cm^3 stoppered conical flask (dry)
- spatula
- vinyl tubing
- glass rod
- 50 cm^3 measuring cylinder
- 25 cm^3 beaker
- 100 cm^3 beaker
- $2 \times 250 \text{ cm}^3$ beakers
- 100 cm^3 volumetric flask
- 25 cm^3 pipette
- pipette filler
- Buchner funnel and flask (dry)
- filter paper
- test-tube
- Bunsen burner
- 2 x sintered glass crucibles no. 4
- weighing bottle
- bowl for ice-water bath
- desiccator
- top pan balance
- UV/visible spectrophotometer and cells
- vacuum source
- analytical balance

Reagents

- iodine
- chlorine gas
- dry caesium bromide
- trichloromethane
- hydrochloric acid (1 mol dm^{-3} and 2 mol dm^{-3})
- sodium iodide
- sodium thiosulfate solution (0.05 mol dm^{-3})
- sodium tetraphenylboron solution
- silver nitrate solution (dilute)
- distilled water
- starch solution
- heptane

Safety

During this experiment gloves should be worn at all times as ICl is very corrosive. If any is spilt, douse it immediately with sodium thiosulfate (preferably) or water. After use, any apparatus contaminated with ICl should be placed in the bucket of sodium thiosulfate solution provided.

Iodine: Harmful by inhalation and in contact with skin.

Chlorine gas: May be fatal if inhaled.

Trichloromethane: Harmful if swallowed. Irritating to skin. Danger of cumulative effects. Carcinogen.

Silver nitrate solution: Corrosive. Can cause permanent damage to eyes.

Heptane: Highly flammable.

Route

A	**Preparation of iodine monochloride**
	Weigh iodine
	React iodine with chlorine
	Reweigh to check appropriate amount of chlorine reacted

↓

B	**Preparation and isolation of the polyhalide**
	Add caesium bromide to iodine monochloride
	Isolate and purify product

↓

C	**Thermal decomposition of the polyhalide**
	Heat polyhalide and observe
	Bubble products of thermal decomposition through heptane and record UV/vis spectrum
	Examine residue after heating

↓

D	**Identification of elements in the polyhalide salt**
	Test for presence of caesium
	Test for presence of chlorine, bromine and iodine

↓

E	**Determination of stoichiometry of the polyhalide**
	Quantitative analysis for caesium
	Quantitative analysis for iodine

A   Preparation of iodine monochloride

Iodine monochloride is prepared by passing chlorine through a known mass of iodine until the mass increases by an amount corresponding to equimolar amounts of iodine and chlorine.

$$I_{2\,(s)} + Cl_{2\,(g)} \rightleftharpoons 2ICl_{(l)}$$

Use 8 g of iodine to prepare ICl. Calculate the mass of chlorine needed and check this with a demonstrator before proceeding.

Grind up around 9 g of iodine using a dry pestle and mortar. Transfer 8 g into an internal-seal trap. This may be easiest to do using a small funnel attached to the side arm of the trap with a short piece of tubing. Weigh the trap on a rough balance, and work out what the mass will be when all the iodine has been converted to ICl. Pass dry chlorine into the trap slowly, while shaking or swirling it gently. Check the mass every 30 seconds, until the required mass is reached. If too much chlorine is added, the reaction will continue to give iodine trichloride. Pour the product carefully into a dry, pre-weighed, 25 cm^3 conical flask and stopper the flask.

Record the electronic spectrum of the product in solution in dry heptane. Compare the spectrum of ICl with spectra of chlorine, bromine and iodine in solution. Which of these spectra does the spectrum of your product resemble?

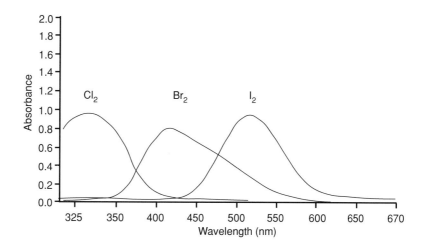

B Preparation and isolation of the polyhalide

In this part of the experiment caesium bromide, CsBr, is reacted with excess ICl. This reaction could have a variety of products. Write down possible products of the reaction.

Take 2 g of finely powdered CsBr, and pour it slowly into the ICl. Stir the mixture thoroughly with a glass rod for two minutes to ensure that all the CsBr has reacted. Add 20 cm^3 of trichloromethane, stir well, stopper the flask and leave to stand for ten minutes, shaking occasionally. The trichloromethane does not react with the product, but precipitates it from solution. Stir thoroughly again to disperse any lumps, and then filter off the crude product using a dry Buchner funnel. Transfer the product to a small beaker, and wash it thoroughly with around five small aliquots of trichloromethane (approximately 5 cm^3), decanting and discarding the washings, until the final wash is almost colourless. Filter again in a dry Buchner funnel. After allowing time for the residual trichloromethane to evaporate, place the product into a dry, stoppered sample tube.

C Thermal decomposition of the polyhalide

Heat a small portion of the polyhalide in a test-tube and record any observations. Place approximately 10 cm^3 of dry heptane in a test-tube. Heat another small portion of the polyhalide as above and dissolve a sample of the gas evolved in the heptane using a Pasteur pipette. Run a UV/visible spectrum of the solution and compare it with the spectrum of ICl and the other halogens.

Dissolve the residue from heating the polyhalide in about 5 cm^3 distilled water and test it with silver nitrate solution and ammonia. Record any observations.

D Identification of elements in the polyhalide salt

The presence of caesium needs to be confirmed. Determine a simple way of doing this. Consult a demonstrator for assistance.

Use observations from part C to identify the halogens in the polyhalide. Check any deductions with a demonstrator.

E   **Determination of stoichiometry of the polyhalide**

Caesium analysis

The proportion of caesium in the polyhalide can be determined quantitatively using gravimetric analysis. This involves taking a known amount of the complex in solution, treating it with a suitable reagent to form an insoluble caesium compound, filtering this off, drying it and weighing it. In this case caesium will be determined gravimetrically as caesium tetraphenylboron, $CsB(C_6H_5)_4$, using a solution of sodium tetraphenylboron as precipitant.

Label two no. 4 sintered glass crucibles using a soft pencil. Place them in the oven at 120 °C to dry. Weigh accurately approximately 0.5 g (but not more than 0.6 g) of the polyhalide and dissolve it in 40 cm^3 of 1 mol dm^{-3} hydrochloric acid solution. Make this up to 100 cm^3 in a volumetric flask using distilled water. Pipette 25 cm^3 of this solution into each of two 250 cm^3 beakers. To each beaker add approximately 0.1 g of sodium iodide and stir until the solid dissolves, then add 20 cm^3 of 0.05 mol dm^{-3} sodium thiosulfate solution. If the solution does not clear completely, add more thiosulfate solution until it does.

Add 35 cm^3 of distilled water to each beaker and then place them in an ice-water bath for about 10 minutes. Add 40 cm^3 of sodium tetraphenylboron solution slowly, with stirring. Leave the beakers in the ice-water bath for 20–25 minutes (but not longer than an hour). While the beakers are cooling, remove the crucibles from the oven and place them in a desiccator to cool, then weigh them accurately. Filter the contents of the beakers, one through each weighed sintered glass crucible, making sure that all the precipitate is transferred. Wash the precipitate several times with small amounts of water, and dry at 100–120 °C for an hour. Cool in a desiccator and reweigh.

Calculate the mass of caesium in the weighed polyhalide sample and then the percent by mass of caesium.

Iodine analysis

The proportion of iodine can be determined quantitatively by titration of the polyhalide with sodium thiosulfate.

Place an accurately weighed sample of the polyhalide (approximately 0.2 g) in a 250 cm^3 conical flask, and dissolve in water. Add an excess (approximately 0.5 g) of sodium iodide, and acidify the solution with about 5 cm^3 of 2.0 mol dm^{-3} hydrochloric acid. Titrate the released iodine immediately with standardised sodium thiosulfate solution (approximately 0.05 mol dm^{-3}).

Sodium thiosulfate solution should be run into the brown solution until it becomes straw coloured, at which point a couple of drops of starch indicator should be added. Continue titrating until the solution becomes colourless. The same volume of starch indicator should be added to each titration.

Polyhalides

This activity has three main stages. Firstly, iodine monochloride, ICl, is prepared by reacting iodine with chlorine. This compound is then treated with caesium bromide to form a polyhalide salt. Finally, the polyhalide is identified and analysed.

Safety

ICl is very corrosive, so gloves should be worn at all times. If any ICl is spilt, douse it immediately with sodium thiosulfate solution (preferably) or water. After use, any apparatus contaminated with ICl should be placed into the bucket of sodium thiosulfate solution provided.

Procedure

From 8.0 g of iodine, 2.25 g of chlorine yields almost quantitatively 10.25 g of iodine monochloride.

Although student activity is relatively prescribed in this experiment, students are not given explicit instructions for procedures such as titration and preparation of solutions. It is assumed that they understand these techniques. There are two main areas where it is anticipated that students will require help or reassurance – when calculating the mass of chlorine that needs to be added to prepare ICl, and before carrying out the practical work required in the thermal decomposition and identification of the polyhalide.

The polyhalide is $CsICl_2$. The presence of caesium can be confirmed by a flame test.

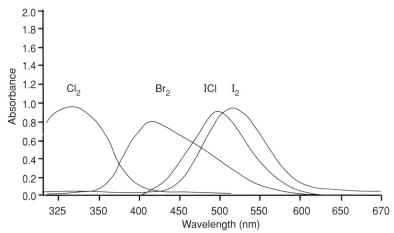

Electronic spectra of chlorine, bromine, iodine and iodine monochloride in CCl_4 solution

T17

Polyhalides

Equipment

- internal-seal trap (dry)
- beaker to support trap
- glass funnel (dry)
- pestle and mortar (dry)
- weighing bottle (dry)
- 25 cm^3 stoppered conical flask (dry)
- spatula
- vinyl tubing
- glass rod
- 50 cm^3 measuring cylinder
- 25 cm^3 beaker
- 100 cm^3 beaker
- 2 x 250 cm^3 beakers
- 100 cm^3 volumetric flask
- 25 cm^3 pipette
- pipette filler
- Buchner funnel and flask (dry)
- filter paper
- test-tube
- Bunsen burner
- 2 x sintered glass crucibles no. 4
- weighing bottle
- bowl for ice-water bath
- desiccator
- top pan balance
- UV/visible spectrophotometer and cells
- vacuum source
- analytical balance

Reagents

- iodine

- chlorine gas

- dry caesium bromide

- trichloromethane

- hydrochloric acid (1 mol dm^{-3} and 2 mol dm^{-3})

- sodium iodide

- sodium thiosulfate solution (0.05 mol dm^{-3})

- sodium tetraphenylboron solution

- silver nitrate solution (dilute)

- distilled water

- starch solution

- heptane

A bucket containing sodium thiosulfate solution (0.05 mol dm^{-3}) should be provided for all apparatus contaminated with ICl.

18. Electrochemical cells

The outcome of this activity depends on a group of students functioning as a team. There are two areas of investigation which together build up a picture of how electrochemical cells operate. These are the concentration dependence of cell electromotive forces (emfs) and the measurement of the electrode potential for the $Co^{2+}|Co$ system.

This experiment is suitable for:

- second year students

- approximately five hours

- group work

Activity type

formal	○	experimental	○	divergent	●	investigatory	○

Skills

manipulation	weighing	○	volume	●	handling	●						
techniques	reflux/distillation	○	recrystallisation	○	chromatography	○	inert atmosphere	○	spectroscopy	●	titration	●
observation	colour	○	volume	●	temperature	○	pressure	○	physical state	●		
data collection	qualitative	●	numerical	●	spectral	○	electronic	○				
data processing	calculation	●	computing	●	matching	○						
interpretation	selection	●	validation	○	deduction	●	prediction	○				
problem solving	identification	●	in/output	○	breakdown	○	methods	●	assembly	○		
team work	skills identification	●	analysis	○	role assignment	○	organisation	○	resources	●	outcomes	●
experiment design	input	●	output	●	precision	○	techniques	●	validity	○		
communication	report	●	poster	○	oral	○	audience	○				
safety	COSHH	○	application	●	review	○	disposal	○				

RS•C
PROGRESSIVE
DEVELOPMENT
OF PRACTICAL
SKILLS IN
CHEMISTRY

137

S18

Electrochemical cells

An electrochemical cell is a device for bringing about a spontaneous chemical reaction in such a way that complementary electron-transfer processes (oxidation and reduction) take place at separated sites. Both the polarity of a cell and the electromotive force (emf) are important characteristics of the system. In this experiment the emfs of a variety of cells are measured, and the way in which such measurements depend on the concentrations of the active cell components are examined.

Equipment

- saturated calomel electrode
- platinum-blacked electrode
- carbon rod
- digital voltmeter (DVM)
- salt bridges
- 2 x 250 cm^3 beakers
- 2 x clamps and retort stands
- 25 cm^3 pipette
- 250 cm^3 volumetric flask
- 6 x 500 cm^3 beakers
- thermometer
- magnetic stirrer
- hydrogen cylinder

Reagents

- sodium chloride solution (0.1 mol dm^{-3} and 0.3 mol dm^{-3})
- sodium hydroxide solution (0.1 mol dm^{-3} and 0.01 mol dm^{-3})
- hydrochloric acid (0.1 mol dm^{-3} and 0.01 mol dm^{-3})
- buffer solutions at pH ~11, 9, 7 and 4
- cobalt(II) sulfate solution (0.1 mol dm^{-3})
- cobalt wire
- cobalt plating solution
- nitric acid (2 mol dm^{-3})
- zinc(II) sulfate solution (0.1 mol dm^{-3})
- copper sulfate solution (0.1 mol dm^{-3})
- distilled water

- silver nitrate solution (0.1 mol dm^{-3})

- ammonia solution (specific gravity 0.880)

- silver electrode

- high purity platinum wire

Safety

Hydrogen: Highly flammable. Explosive mixtures with air.

Sodium hydroxide solutions: Corrosive. Causes severe burns. Risk of permanent damage to eyes.

Hydrochloric acid solutions: Causes burns. Irritating to eyes, skin and respiratory system.

Cobalt(II) sulfate solution: Harmful by inhalation and if swallowed. Irritating to skin, eyes and respiratory system. May cause sensitisation by skin contact.

Nitric acid: Corrosive. May cause eye or skin burns if contact occurs.

Zinc(II) sulfate solution: Irritating to eyes and skin.

Copper sulfate solution: Harmful if swallowed. Irritating to skin and eyes.

Silver nitrate solution: Irritates eyes and causes burns.

Ammonia solution: Corrosive. Irritating to eyes, respiratory system and skin.

Saturated potassium chloride solution: May be irritating to eyes.

Part 1: Concentration-dependence of the cell emf

Experiment A
This experiment is concerned with the cell represented by the following cell diagram:

$$Hg_{(l)}|Hg_2Cl_{2(s)}|Cl^-_{(sat)}|Cl^-_{(aq)}|AgCl_{(s)}|Ag_{(s)} \qquad (1)$$

in which a saturated calomel electrode is combined with a silver/silver chloride electrode. The latter comprises a rod of silver, electrolytically coated with silver chloride, in contact with a solution of chloride ions of known concentration. The electrode potential of this half-cell depends on the establishment of the following equilibrium at the surface of the electrode:

$$AgCl_{(s)} + e^- \rightleftharpoons Ag_{(s)} + Cl^-_{(aq)} \qquad (2)$$

(a) Follow the instructions in Part 3 to prepare a silver/silver chloride electrode.

(b) Two stock solutions of aqueous sodium chloride with concentrations of 0.1 mol dm^{-3} and 0.3 mol dm^{-3} are provided. Use these solutions to prepare sequentially 250 cm^3 of NaCl solutions of concentrations:

0.01, 0.001 and 0.0001 mol dm^{-3}

and   0.03, 0.003 and 0.0003 mol dm^{-3}

Use a 25 cm^3 pipette, a 250 cm^3 volumetric flask and distilled water. In each sequence, ensure that the first solution prepared is thoroughly mixed, and then store it in a clean, dry and labelled beaker before going on to prepare the next solution.

(c) Make up a cell. Start with the most dilute NaCl solution. Place the beaker on a magnetic stirrer, and insert a calomel electrode (making sure it is clean and dry) and your silver/silver chloride electrode. Record the polarity of the cell and its emf, E_A, when it is constant to within 2–3 mV (it should be about 200 mV). Note the temperature of the solution.

(d) Repeat the measurements in (c) with NaCl solutions of increasing concentration starting with the 0.0003 mol dm^{-3} solution and finishing with the stock 0.1 mol dm^{-3} solution. Do not use the 0.3 mol dm^{-3} solution. Take care to minimise cross-contamination of the different solutions.

(e) Record the results as a table of values of E_A (the emf of the cell as represented by the cell diagram) against the concentration of Cl$^-$ in the NaCl solution, $c(\text{Cl}^-)$.

Experiment B

This experiment is concerned with the cell represented by the following cell diagram:

$$\text{Hg}_{(l)}|\text{Hg}_2\text{Cl}_{2(s)}|\text{Cl}^-_{(sat)}|\text{H}^+_{(aq)}|\text{H}_{2(g)},\text{Pt}_{(s)} \qquad (3)$$

in which a saturated calomel electrode is combined with a hydrogen electrode. The latter is a device designed to ensure rapid attainment of equilibrium in the following reaction:

$$\text{H}^+_{(aq)} + \text{e}^- \rightleftharpoons \tfrac{1}{2}\,\text{H}_{2(g)} \qquad (4)$$

The form of this electrode causes hydrogen gas at atmospheric pressure to bubble over platinum metal immersed in an aqueous solution of known pH. The platinum is 'blacked' to give maximum surface area and to catalyse attainment of the equilibrium in equation 4. The gas then bubbles through a few millimetres depth of water before escaping to the atmosphere, as shown in the diagram, thus ensuring that the solution under test is in contact with hydrogen, and minimising back-diffusion of air.

The platinum-blacked electrode has been prepared. In addition, the following stock solutions are provided:

■ NaOH of concentrations 0.1 and 0.01 mol dm^{-3};

■ HCl of concentrations 0.1 and 0.01 mol dm^{-3}; and

■ buffer solutions of pH ~ 11, 9, 7 and 3.

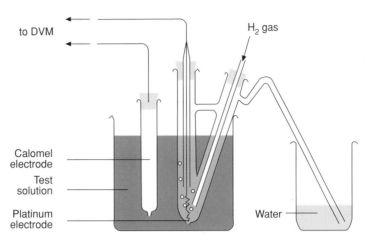

to DVM

H₂ gas

Calomel
electrode

Test
solution

Platinum
electrode

Water

A schematic picture of the experimental set-up

(a) Set up a cell as indicated in the figure, starting with the most concentrated NaOH solution.

(b) Consult a demonstrator if you are inexperienced in the use of a hydrogen cylinder. Adjust the gas flow-rate to 1–2 bubbles per second over the platinum electrode. Leave the system for 12–15 minutes for the H_2 gas to saturate the test solution, and for the equilibrium in equation 4 to be established.

(c) When the potential is constant to within 2–3 mV, record the polarity of the cell and its emf, E_B (which should be in the range 975–1000 mV with 0.1 mol dm^{-3} NaOH).

(d) Repeat the procedure in (b) and (c) with the remaining solutions. Ensure that the solution 'inside' the hydrogen electrode is representative of the solution under test, and take care to minimise cross-contamination of the different solutions. Record the atmospheric pressure and the temperature of the solutions at least once during the series of measurements. Ensure that the hydrogen cylinder is properly closed down at the end of the experiment – consult a demonstrator if necessary.

(e) Record the results as a table of values of E_B (the emf of the cell as represented by cell diagram 3) against the pH of the solution under test.

Data analysis (for experiments A and B)

(a) Plot a graph of E(cell) against $\log (c/c^{\ominus})$, where $c^{\ominus}$ is the standard concentration ($c^{\ominus} = 1$ mol dm^{-3}) and c is the concentration of either $Cl^{-}_{(aq)}$ (experiment A) or $H^{+}_{(aq)}$ (experiment B).

(b) The plot should be a straight line – within experimental error. Read off the intercept, and determine the magnitude, sign and unit of the slope. Hence summarise the experimental measurements as an equation of the form:

$$E(\text{cell}) = (\text{intercept}) + (\text{slope}) \log (c/c^{\ominus}) \tag{5}$$

PROGRESSIVE
DEVELOPMENT
OF PRACTICAL
SKILLS IN
CHEMISTRY

■

Part 2: Measurement of the electrode potential for the Co^{2+} | Co system

This part is designed to highlight some of the many practical problems involved in the direct determination of electrode potentials that are sufficiently accurate to be incorporated in the chemical literature.

Consider the measurement of $E(Co^{2+}|Co)$, the electrode potential for the $Co^{2+}|Co$ system. In addition to the materials supplied in earlier parts of the electrochemistry experiment, a stock solution of $CoSO_4$ of concentration 0.1 mol dm^{-3} and a sample of cobalt wire are provided.

The aim of the experimental work is to identify some of the factors that affect the value of $E(Co^{2+}|Co)$ measured relative to the saturated calomel electrode. In order to do this, use the cell represented by the following cell diagram:

$$Hg_{(l)}|Hg_2Cl_{2(s)}|Cl^-_{(sat)}|Co^{2+}_{(aq)}; 0.1 \text{ mol } dm^{-3}|Co_{(s)} \tag{6}$$

Start by calculating the cell emf (at 298.15 K), using the following literature values of the electrode potentials (on the standard hydrogen electrode scale):

E(sat. calomel) $= + 0.224$ V

$E^{\ominus} (Co^{2+}|Co)$ $= - 0.28$ V

Note down any assumptions made. Then set up the cell and measure its emf.

(a) Prepare your cobalt wire electrode by cleaning it in nitric acid (2 mol dm^{-3}). Set up a cell comprising a cobalt electrode immersed in a 0.1 mol dm^{-3} solution of $CoSO_4$ and a saturated calomel electrode in saturated potassium chloride. Record the cell polarity and its emf.

(b) Note the effect of adding two drops of each of the solutions (a)–(d), in the order given below. Make a note of any observations.

  (a) distilled water

  (b) 0.1 mol dm^{-3} $ZnSO_4$

  (c) 0.1 mol dm^{-3} $CuSO_4$

  (d) concentrated HNO_3 (2 mol dm^{-3})

Discussion

Gather together as a group and compare results. Try to account for the observed dependence of emf on the method of preparation of the cobalt electrode and the effect of adding various solutions. A published account of the measurement of the electrode potential of nickel may be useful (*J. Phys. Chem.*, 1929, **33**, 161–178). Extract any important information relevant to the determination of $E^{\ominus}$ for cobalt from this publication.

RS•C

Part 3: Preparation and storage of electrodes

Saturated calomel electrode
The saturated calomel electrode is provided ready for use. Wash it thoroughly with distilled water before placing it in a solution. Store the electrode in the saturated solution of potassium chloride.

Silver/silver chloride electrode
Take a silver electrode (previously coated with silver chloride) and clean it in concentrated ammonia solution (specific gravity 0.880) for about 2 minutes in a fume cupboard. Rinse the electrode with distilled water, and then place it in a concentrated solution of nitric acid until gassing commences (about 30 seconds). Wash the electrode with distilled water and then place it in a $0.1 \ mol \ dm^{-3}$ solution of hydrochloric acid. Set up the appropriate circuitry to electrolyse the solution. The silver electrode forms the anode and a pure platinum wire electrode forms the cathode (be careful to select the correct platinum electrode).

Use a digital voltmeter (DVM) to measure the current. Make sure the potential is set to zero and then switch on. Increase the potential difference across the system until the current has a value of approximately 1.8 mA, and continue the electrolysis until a uniform coating of silver chloride is formed on the silver electrode. This should take approximately 30 minutes. Store the electrode in sodium chloride solution ($0.1 \ mol \ dm^{-3}$) for at least one hour before use. Before placing the electrode in more dilute sodium chloride solutions, wash it thoroughly with distilled water. When the experiment is complete, store the electrode in distilled water.

Hydrogen electrode
The platinum-blacked electrode that forms part of the hydrogen electrode has been prepared in advance. It is stored in distilled water.

Cobalt electrode
Suspend a carbon rod and a pure platinum wire electrode in a small beaker containing the 'cobalt plating solution' (containing $Co^{2+}_{(aq)}$ ions) provided. Set up the appropriate circuitry to electrolyse the solution, and hence plate cobalt metal onto the carbon rod. Use a DVM to measure the current. Make sure the potential is set to zero and then switch on. Increase the potential difference across the system until the current has a value of approximately 200 mA, and continue the electrolysis until a uniform coating of cobalt metal is formed on the carbon rod. This should take about 20 minutes. Store the electrode in $0.1 \ mol \ dm^{-3}$ cobalt sulfate solution for as long as possible before use.

D18

Electrochemical cells

In this experiment the emfs of a variety of cells are measured, and the way in which such measurements depend on the concentrations of the active cell components are examined.

In part one, students should be grouped in pairs, with each person carrying out either experiment A or B. The analysis can be done together, and differences between the two systems compared. For part two, it is best to organise students into groups of about six. They can report their conclusions in writing or by giving a short talk.

Students may need to be shown how to use the digital voltmeter (DVM), although most will have had experience with such equipment in school.

Part 1: Concentration-dependence of the cell emf

Experiment A

Each pair of students is provided with only two volumetric flasks, so they may need to be encouraged to prepare each series of solutions sequentially, storing each one in a labelled beaker. The final solutions are very dilute, so flasks must be washed thoroughly each time – likewise the beaker used for the cell, and the electrodes themselves. Stirring is also important.

Experiment B

Care should be taken to ensure that students know how to use the hydrogen cylinder, and that it is closed down properly when they have finished. If the emf reading takes longer than 30 minutes to become steady, it is likely that the platinum-blacked electrode is unsatisfactory.

Part 2: Measurement of the electrode potential for the Co^{2+} | Co system

Calculation of the cell emf

E(cell) $= E_{RHE} - E_{LHE}$

$\qquad = E(Co^{2+}|Co) - E(\text{sat. calomel})$

$E(Co^{2+}|Co) = E^{\ominus}(Co^{2+}|Co) - (RT/2F) \ln\{a(Co)/a(Co^{2+})\}$

$\qquad = E^{\ominus}(Co^{2+}|Co) + (RT/2F) \ln a(Co^{2+}), \text{ since } a(Co) = 1$

Assuming ideal behaviour, set $a(Co^{2+}) \sim c(Co^{2+})/c^{\ominus} = 0.1$

Then

$E^{\ominus}(Co^{2+}|Co) = \{-0.28-0.03\}V = -0.31 \text{ V}$

so

E(cell) $= \{-0.31-0.244\}V = -0.554 \text{ V}$

Assuming $\gamma \pm$ (0.1 M $CoSO_4$) $= 0.155$ (as for $CuSO_4$ and $ZnSO_4$)

Then

$E^{\ominus}(Co^{2+}|Co) = \{-0.28-0.05\}V = -0.33 \text{ V}$

so

E(cell) = {-0.33-0.244} V = -0.574 V

(a) Typical results (although not very consistent) for the cell emf are:

    cleaned Co metal: -0.47 V

    Co metal + Hg: -0.48 V (for information)

    Co metal on Pt: -0.51 V (for information)

    Co metal on C: -0.52 V

(b) Distilled water   - hardly any effect (drop up to 5 mV)

    0.1 M $ZnSO_4$ - not much different from distilled water
           (drop up to 10 mV)

    0.1 M $CuSO_4$ - large effect, potential falls by ~0.3 V

    Nitric acid   - large effect, potential falls by ~0.4 V

It should take around three hours for students to reach this point.

Students should pick out the following as important points relevant to the determination of $E^{\ominus}($ Co^{2+}|Co):

■ The temperature must be controlled.

■ The materials used must be pure.

■ Oxygen gas must be excluded.

■ Hydrogen gas should be excluded.

■ Oxygen produces electrode potential values that are too positive; hydrogen produces values that are too negative.

■ Values must be corrected for activity coefficients.

■ Liquid junction potentials should be removed (*eg* by using a single electrolyte) or taken into account.

■ Nickel wire is likely to have strain associated with it; measurements should be made with finely divided nickel.

■ Nickel can displace hydrogen from an acidic solution. This hydrogen will affect the results and it is therefore necessary to work at a near neutral pH. At higher pH, complications from hydroxide arise.

■ It is important to remove impurities especially those more electropositive than Ni, *eg* Cu. Such impurities can lead to more positive electrode potential values.

■ The cells need to be reversible.

■ The suggestion is made that either cobalt is more positive than nickel (not so) or it shows less cathodic polarisation than nickel (*ie* it has a higher rate of deposition than nickel).

Observations on cobalt electrode system

The value of the electrode potential produced is dependent on the method of preparation of the metal electrode. Pure cobalt wire is likely to have an electrode potential 0.1 V more positive than the electrode chosen as producing the most reproducible and reversible potentials. The nickel amalgam electrodes are similar to pure nickel wire, a similar result being observed for cobalt. The electrolytically produced nickel and cobalt electrodes showed consistently higher negative values of the electrode potential.

The following paper – *J. Phys. Chem.*, 1929, **33**, 161 – suggests that the most appropriate electrode to select would be an electrolytically produced cobalt electrode (on platinum, for example) completely covered by a sample of finely divided cobalt metal (produced electrolytically and knocked off the electrode), prepared in an oxygen environment. Such an electrode is said to be in its most stable state, and free from oxide.

Most students would not expect the electrode potential of cobalt to be affected by the addition of copper or hydrogen ions to the system. The explanation for changes which occur can be found, to some extent within the article above.

T18

Electrochemical cells

Equipment

- saturated calomel electrode
- platinum-blacked electrode
- carbon rod
- digital voltmeter (DVM)
- salt bridges
- 2 x 250 cm^3 beakers
- 2 x clamps and retort stands
- 25 cm^3 pipette
- 250 cm^3 volumetric flask
- 6 x 500 cm^3 beakers
- thermometer
- magnetic stirrer
- hydrogen cylinder

Reagents

- sodium chloride solution (0.1 mol dm^{-3} and 0.3 mol dm^{-3})
- sodium hydroxide solution (0.1 mol dm^{-3} and 0.01 mol dm^{-3})
- hydrochloric acid (0.1 mol dm^{-3} and 0.01 mol dm^{-3})
- buffer solutions at pH ~11, 9, 7 and 4
- cobalt sulfate solution (0.1 mol dm^{-3})
- cobalt wire
- cobalt plating solution
- nitric acid (2 mol dm^{-3})
- zinc sulfate solution (0.1 mol dm^{-3})
- copper sulfate solution (0.1 mol dm^{-3})
- distilled water
- silver nitrate solution (0.1 mol dm^{-3})
- ammonia solution (specific gravity 0.880)
- silver electrode
- high purity platinum wire

LIVERPOOL JOHN MOORES UNIVERSITY
LEARNING SERVICES

19. Unknown alcohol

This is primarily an exercise in experimental planning and searching scientific literature. The object of the experiment is to identify an unknown secondary alcohol by converting it to the corresponding ketone, preparing a 2,4-dinitrophenylhydrazine derivative, and identifying the derivative by melting point determination. There are no experimental details given in the student guide, so it is necessary to plan the experimental procedure using information found by searching through literature.

This experiment is suitable for students approaching the end of the first year, or in the second year. It is anticipated they would need at least two weeks to research the experiment and produce a written procedure, depending on commitments from other parts of their degree course. It should be possible to complete the practical work in three to four hours.

This experiment is suitable for:

■ end of first year/second year students

■ experimental research and design – two weeks

■ practical work – three to four hours

■ pairs/groups

Activity type

formal	○	experimental	●	divergent	○	investigatory	○

Skills

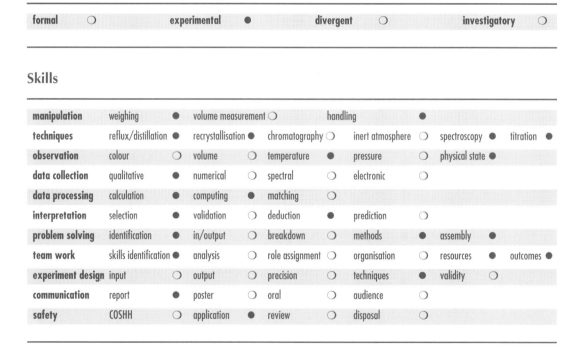

manipulation	weighing ●	volume measurement ○	handling ●		
techniques	reflux/distillation ●	recrystallisation ●	chromatography ○	inert atmosphere ○	spectroscopy ●	titration ●
observation	colour ○	volume ○	temperature ●	pressure ○	physical state ●
data collection	qualitative ●	numerical ○	spectral ○	electronic ○	
data processing	calculation ●	computing ●	matching ○		
interpretation	selection ●	validation ○	deduction ●	prediction ○	
problem solving	identification ●	in/output ○	breakdown ○	methods ●	assembly ●
team work	skills identification ●	analysis ○	role assignment ○	organisation ○	resources ●	outcomes ●
experiment design	input ○	output ○	precision ○	techniques ●	validity ○
communication	report ●	poster ○	oral ○	audience ○	
safety	COSHH ○	application ●	review ○	disposal ○	

Unknown alcohol

This activity introduces the use of literature for finding experimental procedures. Having found a suitable procedure it may be necessary to adapt it before carrying it out in the laboratory. A detailed procedure must therefore be determined for this experiment, in which an unknown alcohol must be identified.

Equipment

- 100 cm^3 measuring cylinder
- 3 x 50 cm^3 beakers
- 2 x 100 cm^3 beakers
- 100 cm^3 conical flask
- 250 cm^3 conical flask
- 250 cm^3 round-bottomed flask
- funnel and filter paper
- thermometer
- ice bath
- anti-bumping granules
- distillation equipment
- 250 cm^3 separating funnel
- spatula
- glass rod
- Buchner funnel
- top pan balance
- melting point apparatus

Reagents

- sodium dichromate dihydrate
- concentrated sulfuric acid
- oxalic acid
- sodium chloride
- dichloromethane
- ethanol
- 2,4-dinitrophenylhydrazine

Unknown alcohols

■ butan-2-ol

■ pentan-2-ol

■ cyclopentanol

■ cyclohexanol

■ cycloheptanol

■ 2-methylcyclohexanol

Safety

Sodium dichromate dihydrate: Very toxic by inhalation. Harmful in contact with skin.

Concentrated sulfuric acid: Burns skin and eyes. Causes severe damage if taken by mouth. Toxic by inhalation of fumes or mist.

Oxalic acid: Irritating to eyes and respiratory system. Harmful in contact with skin or if swallowed.

Dichloromethane: Highly volatile and flammable. Will pressurise a closed vessel when shaken. Will degrease skin on contact. Harmful by inhalation. Irritating to eyes. Toxic if taken by mouth.

Ethanol: Harmful by inhalation or ingestion. Highly flammable.

2,4-Dinitrophenylhydrazine: Irritant to skin and eyes. Harmful if inhaled, ingested or absorbed through skin.

Alcohols: Flammable. Irritating to eyes, skin and respiratory system. Harmful by inhalation and ingestion.

A small volume (around 50 cm^3) of a secondary alcohol is provided. The purpose of this experiment is to identify it by first oxidising it to a ketone, and then preparing a 2,4-dinitrophenylhydrazine (DNP) derivative of the ketone. DNP derivatives of ketones and aldehydes are usually crystalline solids, and have very distinctive melting points which depend on the ketone. By recording this melting point, together with the boiling point of the ketone, it is possible to identify the starting alcohol.

Before arriving in the laboratory, it is necessary to prepare a written plan of exactly how this work is to be carried out. It must contain details about quantities and concentrations of reagents, as well as hazards associated with each reagent and technique, and precautions that need to be taken. The experiment should be scaled so that approximately 10 g of the unknown alcohol is used. A demonstrator must be shown the procedure before work starts. In the event that the procedure needs altering (which is more likely to be due to availability of reagents and equipment than because it is a bad procedure), a demonstrator will provide guidance. A demonstrator will also instruct you on how to perform a COSHH assessment.

Start the literature search with a standard practical organic textbook such as L. M. Harwood and C. J. Moody, *Experimental Organic Chemistry: Principles and Practice*, Oxford, Blackwell Scientific, 1990.

Reference your work fully, as this will make it easier to make any necessary amendments, and is good scientific practice.

D19

Unknown alcohol

This experiment is designed primarily to give students experience of searching through literature, and planning their own experiments. The experiment is fairly standard, however students have not been given a procedure to work from. Instead, they need to read about the subject and produce a written plan of how they would carry out the experiment. This experiment is particularly suitable for group or pair work.

Students have the task of oxidising an unknown secondary alcohol to a ketone, and then preparing a 2,4-dinitrophenylhydrazine (DNP) derivative of this ketone. The melting point of the DNP derivative, together with the boiling point of the ketone can be used to identify the ketone, and hence the starting alcohol. Students have also been given an indication of the scale to work on, using about 10 g of their starting alcohol.

At least two weeks are needed in which to research this experiment. Marked scripts could be returned, or work discussed with students a week before the practical laboratory session. This allows them time to process any comments and make corrections accordingly.

Checking students' procedures

This section is intended to be a guide to some of the relevant issues that are involved when checking students' completed procedures. It is not a guide in how to teach or mark work, but merely suggestions that have arisen during the trialling of this experiment.

1   Is the procedure chemically viable?

It is anticipated that most procedures presented would be viable. However, students may merely think about how the reaction is presented in lecture notes, and have little understanding about catalysts, conditions or rates of reaction. They may not fully understand what working up a reaction means, and this would be obvious in their procedure.

Students may present a procedure that will work, even though a faster process or one which produces a higher yield is known. In this case, within reason, students should be allowed to follow their own procedures. However, if the suggestion is totally unreasonable, they can be reassured that there was nothing wrong with their work and that there is simply a better method.

2   Are the relevant safety instructions included?

The importance of proper safety precautions must be made clear to the student. In most cases, students will have tried to save time by not considering safety aspects fully. They may have had trouble finding the information or may not fully understand its importance.

3   Are all the reagents and equipment necessary for the students to carry out their procedures available?

It is expected that most students will use a procedure similar to the one outlined below. However, some procedures may use different chemicals.

If scripts are collected early enough, it should be possible to assemble all of the compounds and solutions required, although this is a very costly approach.

The procedure below can be given to those students whose suggestions were not viable due to lack of resources. However, it is important to indicate to students that this is because of a problem with resources and not because they have presented a bad procedure.

Sample procedure

Alcohol oxidation

Prepare an acidified sodium dichromate solution (the oxidising agent) by dissolving 20 g of sodium dichromate in 100 cm^3 of distilled water. While stirring, carefully add 10 cm^3 of concentrated sulfuric acid. Cool the solution by placing it in cold water for a few minutes.

Weigh 10 g of the alcohol into a 250 cm^3 round-bottomed flask. Add all of the acidified sodium dichromate solution in one portion, and swirl. The mixture will become hot. The temperature of the mixture must be kept between 55 and 60 °C by cooling the flask in cold water when necessary. When the temperature of the mixture ceases to rise when removed from the cold water, allow it to stand with occasional shaking for 20 minutes. Add 2 g of oxalic acid, whilst stirring, to destroy excess dichromate.

Separating and purifying ketone

Add 75 cm^3 of water and a few anti-bumping granules to your reaction mixture. Distil the mixture until around 60 cm^3 of distillate has collected. Saturate the distillate with sodium chloride, and use a separating funnel to separate the organic (upper) layer from the aqueous (lower) layer. Extract the aqueous layer using approximately 35 cm^3 of dichloromethane, and add this organic extract (lower layer) to the original organic extract.

Use anhydrous sodium sulfate to dry the organic layers, then filter off the drying agent. Add a few anti-bumping granules, and remove the solvent by distillation. Distil the residual liquid, recording the temperature at which the main fraction boils (between 80 and 190 °C).

Preparing the 2,4-dinitrophenylhydrazone derivative

To prepare the 2,4-dinitrophenylhydrazine reagent, dissolve 1 g of 2,4-dinitrophenylhydrazine in 20 cm^3 of ethanol. Add 2 cm^3 of concentrated sulfuric acid and filter under gravity if necessary. Dissolve 0.5 cm^3 of your product in 10 cm^3 of ethanol. Add the solution of 2,4-dinitrophenylhydrazine reagent in ethanolic sulfuric acid and allow the mixture to stand at room temperature to crystallise. When crystallisation is complete, filter off the 2,4-dinitrophenylhydrazone derivative and wash with a little ice-cold ethanol.

Purifying the 2,4-dinitrophenylhydrazone derivative

Recrystallise with 95% ethanol.

Identifying the alcohol

To identify the starting alcohol, the melting point of the
2,4-dinitrophenylhydrazone derivative must be measured. Use this, together
with the boiling point of your ketone, to identify the ketone and hence the
starting alcohol.

Alcohol	Boiling point of corresponding ketone/°C	Melting point of corresponding DNP derivative/°C
butan-2-ol	80	111
pentan-2-ol	102	144
cyclopentanol	131	147
cyclohexanol	155	162
cycloheptanol	181	148
2-methyl-cyclohexanol	163	137

Unknown alcohol

Equipment

- 100 cm^3 measuring cylinder
- 3 x 50 cm^3 beakers
- 2 x 100 cm^3 beakers
- 100 cm^3 conical flask
- 250 cm^3 conical flask
- 250 cm^3 round-bottomed flask
- funnel and filter paper
- thermometer
- ice bath
- anti-bumping granules
- distillation equipment
- 250 cm^3 separating funnel
- spatula
- glass rod
- Buchner funnel
- top pan balance
- melting point apparatus

Reagents

- sodium dichromate dihydrate
- concentrated sulfuric acid
- oxalic acid
- sodium chloride
- dichloromethane
- ethanol
- 2,4-dinitrophenylhydrazine (DNP)

Unknown alcohols

■ butan-2-ol

■ pentan-2-ol

■ cyclopentanol

■ cyclohexanol

■ cycloheptanol

■ 2-methylcyclohexanol

Additional reagents may be required if alternative procedures, as described in the student notes, are followed.

20. Equilibrium constant

This experiment investigates the dimerisation equilibrium constant between N_2O_4 and NO_2. Students are provided with the opportunity to learn how to use vacuum lines – a technique that has uses in many different areas of chemistry.

The idea for this experiment came from a laboratory book kindly sent to us by the University of Cambridge.

This experiment is suitable for:

■ first year students

■ approximately three hours

■ individuals/pairs

Activity type

| formal | ● | experimental | ○ | divergent | ○ | investigatory | ○ |

Skills

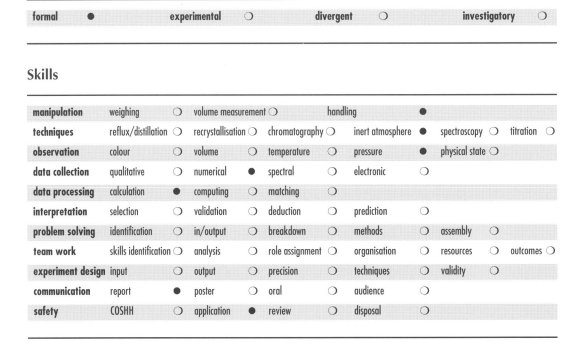

manipulation	weighing	○	volume measurement	○	handling	●			
techniques	reflux/distillation	○	recrystallisation ○	chromatography ○	inert atmosphere	●	spectroscopy ○	titration ○
observation	colour	○	volume	○	temperature ○	pressure	●	physical state ○
data collection	qualitative	○	numerical	●	spectral ○	electronic	○	
data processing	calculation	●	computing	○	matching ○			
interpretation	selection	○	validation	○	deduction ○	prediction	○	
problem solving	identification	○	in/output	○	breakdown ○	methods	○	assembly ○
team work	skills identification ○	analysis	○	role assignment ○	organisation	○	resources ○	outcomes ○
experiment design	input	○	output	○	precision ○	techniques	○	validity ○
communication	report	●	poster	○	oral ○	audience	○	
safety	COSHH	○	application	●	review ○	disposal	○	

S20

Equilibrium constant

In this activity, a value for the equilibrium constant for the dimerisation of NO_2 to N_2O_4 is measured. This is achieved by measuring the pressure of the equilibrium mixture in a closed system at several different temperatures.

Pre-laboratory work

This pre-laboratory session covers two topics – the use of a vacuum line, and the theory behind the experiment.

Using a vacuum line

Using a vacuum line needs careful thought, especially if experience with the equipment is limited. Turning the wrong tap at the wrong time can result in damaging the equipment or releasing dangerous gasses, as well as often necessitating that the experiment be repeated. Before changing the position of a tap think about what you want to do, and whether it is the same as what you are about to do. The questions below have been designed to make you think about the effect of turning each tap. Refer to the diagram of the equipment to answer them.

Tap 1 is used to isolate the left and right hand sides of the manometer. If this tap is open, the two oil levels should be the same. To read the pressure, close tap 1, and measure the difference in the height of the two levels.

1   In order to pump gases from the whole apparatus (*ie* place the system under vacuum), in which position should each tap be?

Tap 1   Open / Closed / Not important

Tap 2   Open / Closed / Not important

Tap 3   Open / Closed / Not important

Tap 4   Open / Closed / Not important

Tap 5   Open / Closed / Not important

Tap 6   Open / Closed / Not important

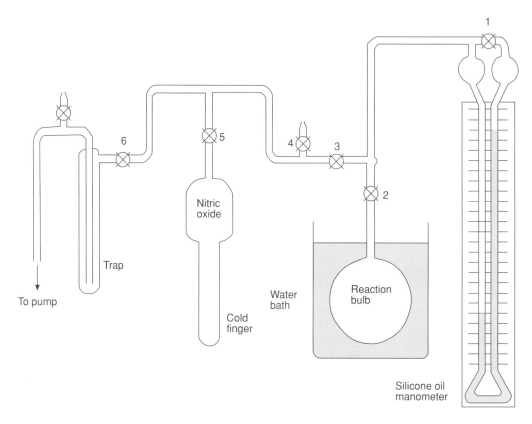

To pump

Trap

Nitric oxide

Cold finger

Water bath

Reaction bulb

Silicone oil manometer

2   To remove gasses from the reaction bulb, but not lose any nitric oxide, in which position should each tap be?

   Tap 1   Open / Closed / Not important

   Tap 2   Open / Closed / Not important

   Tap 3   Open / Closed / Not important

   Tap 4   Open / Closed / Not important

   Tap 5   Open / Closed / Not important

   Tap 6   Open / Closed / Not important

3   The temperature of liquid nitrogen is low enough to freeze N_2O_4. How could the gasses be removed from the entire system without losing a significant amount of N_2O_4?

4   Which other gasses normally present in air will freeze in liquid nitrogen? How can their concentration in the cold finger be reduced?

5   Pump out the system, then add NO_2 to the reaction bulb, and measure the pressure of NO_2 added. In note form, detail the sequence of events followed, and any observations made, before proceeding to the next step.

Theoretical background

The following questions have been designed to introduce some of the theory behind the experiment. Particular attention is paid to the mathematical manipulations that will be carried out on the data.

The equilibrium between NO_2 and N_2O_4 will be studied.

$$N_2O_4 \rightleftharpoons 2NO_2$$

As the dimerisation is exothermic, lowering the temperature will increase the amount of the dimer in the system at equilibrium.

The equilibrium constant can be expressed in terms of partial pressures:

$$K_p = \frac{(pNO_2/p°)^2}{(pN_2O_4/p°)} \tag{1}$$

where $p°$ is the standard pressure, and pNO_2 and pN_2O_4 are the partial pressures of NO_2 and N_2O_4 respectively. The $p°$ terms can be omitted, provided that the partial pressures are expressed in atmospheres.

1 n moles of pure N_2O_4 are introduced into a vessel. If at equilibrium m moles of N_2O_4 have dissociated, how many moles of NO_2 will have been produced? How many moles of N_2O_4 are now left?

2 At equilibrium, what is the total number of moles of gas present (in terms of n and m)?

3 If the volume of the vessel, the total pressure in the vessel, the temperature the vessel is held at, and the value of n are known, how is it possible to calculate m?

Equation (1) above shows that, in order to find the value of K_p the partial pressures of both NO_2 and N_2O_4 must be known. These can be related to the total pressure, p_{tot}, using Dalton's Law, and the mole fractions of each of the gases.

$$p_A = x_A p_{tot} \tag{2}$$

where the mole fraction, x_A, is the ratio of the number of moles of A to the total number of moles of gas present:

$$x_A = \frac{n_A}{n_{tot}} \tag{3}$$

4 Using the values for the total number of moles of gas, and the number of moles of NO_2 and N_2O_4 obtained in questions 1 and 2 above, write expressions for the mole fraction of N_2O_4 and NO_2 (xN_2O_4 and yNO_2 respectively) in terms of n and m.

Because N_2O_4 is a mobile, volatile liquid, it is very hard to admit a known amount directly into the vessel. However, by calibrating the vessel with air, the need to know the precise volume of the vessel is removed.

A calibration curve can be plotted by admitting air into the vessel, closing the tap, and recording the change in pressure with increasing temperature, until the water in the water bath is boiling. The air is then pumped away.

N_2O_4 is admitted to the vessel at an equal pressure to the air at that temperature, while the water bath is kept at 100 °C. At this temperature the N_2O_4 is completely dissociated to NO_2.

5 As the temperature drops, the NO_2 will dimerise. The pressure will drop for two reasons. What are these?

6 If two moles of NO_2 associate to form a mole of N_2O_4, how has the total number of moles in the system changed? Any change of total number of moles will be reflected directly by a change in pressure. How will the difference in pressure between the air and the NO_2/N_2O_4 systems at a given temperature be related to the number of moles (and also the partial pressure in this case) of N_2O_4 in the system?

The total pressure is the sum of all the partial pressures. Therefore, for the N_2O_4/NO_2 equilibrium mixture:

$$p_N = p_{N_2O_4} + p_{NO_2} \tag{4}$$

where p_N is the observed pressure of the N_2O_4/NO_2 mixture at a given temperature.

As $p_{N_2O_4} = p_{air} - p_{N'}$ (where p_{air} is the observed pressure of the air at the same temperature as $p_{N'}$) this expression can be rearranged, giving:

$$p_{NO_2} = 2p_N - p_{air} \tag{5}$$

7 Using this relationship, rewrite equation (1) in terms of p_N and p_{air}.

This expression is only valid if the pressures of air and NO_2 are identical at the highest temperature.

K_p can be related to the standard enthalpy change of the reaction using the equation:

$$\ln K_p = \frac{-\Delta_r H^\circ}{RT} + constant \tag{6}$$

A plot of $\ln K_p$ against $1/T$ should be a straight line with slope $-(\Delta_r H^\circ/R)$.

Safety

All work involving a vacuum system must take place from behind a safety screen. Wear gloves and a face visor.

Procedure

Bring the water in the water bath to boiling point. Close taps 4 and 5, and immerse the cold finger in liquid nitrogen. Evacuate the entire apparatus, excluding the cold finger.

Use the calibration curve provided to read off the pressure $p_{air, init}$ that corresponds to the temperature of the water bath. Evacuate the cold finger by opening tap 5 (make sure the N_2O_4 is still frozen first). Close taps 5 and 6.

Remove the liquid nitrogen from the cold finger. Open tap 5, and let in NO_2 until the pressure is within 1 or 2 mm of $p_{air, init}$. If the pressure is too high then close tap 2, and replace the cold finger in liquid nitrogen. The NO_2 in the line will freeze, resulting in a fall in pressure. Close tap 5, then open tap 2. The new pressure can now be measured. It may be necessary to let in more NO_2, or freeze more into the cold finger, before the correct pressure is reached. Record the temperature and the pressure.

Add cold water to the bath, until the temperature is around 45 °C. Record the pressure and the temperature. Lower the temperature in 5 °C steps until around 0 °C is reached, recording the pressure and temperature at each step.

Immerse the cold finger in liquid nitrogen (if it is not still immersed) and open tap 5. Leave it open for 2 or 3 minutes, to allow all the NO_2 to freeze back into the cold finger. Close tap 5. Evacuate the entire apparatus.

For each temperature, read the value for p_{air} from the calibration chart, and use this with the observed pressure in the reaction vessel, p_N to calculate K_p. Plot a graph of $\ln K_p$ against $1/T$ to determine a value for $\Delta_r H°$.

Equilibrium constant

In this activity, a value for the equilibrium constant for the dimerisation of NO_2 to N_2O_4 is measured. This is achieved by measuring the pressure of the equilibrium mixture in a closed system at several different temperatures. The experiment is carried out using a vacuum line, giving the students experience with working under reduced pressure. Deciding which tap to open will practise students' problem solving skills, as well as giving them valuable experience with vacuum lines. The pre-laboratory questions are designed to ensure that students understand the mathematics behind the experiment, rather than blindly carrying out calculations on their data.

Pre-laboratory work

Use of vacuum line

1 Tap 1   **Open** / Closed / Not important

   Tap 2   **Open** / Closed / Not important

   Tap 3   **Open** / Closed / Not important

   Tap 4   Open / **Closed** / Not important

   Tap 5   **Open** / Closed / Not important

   Tap 6   **Open** / Closed / Not important

2 Tap 1   Open / Closed / **Not important**

   Tap 2   **Open** / Closed / Not important

   Tap 3   **Open** / Closed / Not important

   Tap 4   Open / **Closed** / Not important

   Tap 5   Open / **Closed** / Not important

   Tap 6   **Open** / Closed / Not important

3 By freezing the N_2O_4 in liquid nitrogen before opening taps 5 and 6 and pumping for a few minutes.

4 By freezing the N_2O_4 and pumping as above, then allowing it to defrost and pumping for a few seconds to remove any gasses that have been absorbed onto the surface of the N_2O_4. Repeating this process a few times will remove the majority of contaminants from the line.

5 Freeze N_2O_4 in liquid nitrogen, and open all the taps except 4. Leave for a few minutes to allow the pressure to drop, then close tap 6 and tap 1. Allow the N_2O_4 to defrost, then open tap 5 to allow a small amount into the system. Close tap 5, and read the pressure by measuring the difference between the oil heights on the manometer.

Theoretical background

1 Number of moles of $NO_2 = 2m$
Number of moles of $N_2O_4 = n - m$

2 Total number of moles of gas $= n + m$

3 By using the ideal gas equation, in the form:
$$pV = (n + m)RT$$

4 $xN_2O_4 = \dfrac{(n - m)}{(n + m)}$

$yNO_2 = \dfrac{2m}{n + m}$

5 The pressure will drop because of the change in temperature, but also because for each mole of N_2O_4 formed by dimerisation, the total number of moles will decrease.

6 If two moles of NO_2 associate to form a mole of N_2O_4, the total number of moles in the system will decrease by one. The difference in partial pressure between the air and the NO_2/N_2O_4 systems at the same temperature will be equal to the partial pressure of N_2O_4:

7 $K_p = \dfrac{(2p_N - p_{air})^2}{p_{air} - p_N}$

Procedure

Although students have been given diagrams of the equipment, some of them may need help relating what they see on the diagram to the equipment in front of them. They may also need to be shown how to open and close the taps correctly, how to read the manometer, and how to use liquid nitrogen safely.

Students must be provided with a calibration curve for each set of apparatus. This can be produced by admitting a small volume of air into the apparatus, and recording the pressure over a range of temperatures, between 0 and 100 °C, and plotting a graph of the results. It is probably best done by hand, rather than by computer, so that it is easy for students to relate any value of temperature to the associated pressure.

Further investigations

Use the relationships $\Delta_r G° = -RT\ln K_p$ and $\Delta_r G° = \Delta_r H° - T\Delta_r S°$ to find a value for the entropy.

Ask students to prepare calibration curves themselves.

T20

Equilibrium constant

This activity requires the use of a vacuum line. The precise design is not important provided it is capable of the manipulations described in the diagram in the Student Guide. N_2O_4 and liquid nitrogen are also needed.

21. Stereochemistry of nickel compounds

Four coordinate nickel complexes of the type $[NiX_2(PR_3)_2]$ (where $X = Cl^-$, Br^-, I^- or NCS^- and R = phenyl or cyclohexyl) can adopt either square planar or tetrahedral geometries. This team-focused activity involves synthesising a range of these complexes and determining their geometry as solids and in solution by three different methods – the Evans NMR method, solid state magnetic moment and low frequency infrared. The factors influential in determining the geometry are identified and the question asked as to whether it is possible to design a complex which exhibits both geometries. The strategy, experimental design and subsequent work involve extensive team discussions and practical work.

This experiment is suitable for:

- second year students

- four to six laboratory sessions

- group work

Activity type

formal ○	experimental ●	divergent ●	investigatory ○

Skills

manipulation	weighing ●	volume measurement ●	handling ●		
techniques	reflux/distillation ●	recrystallisation ●	chromatography ○	inert atmosphere ●	spectroscopy ● titration ○
observation	colour ●	volume ○	temperature ○	pressure ○	physical state ●
data collection	qualitative ●	numerical ●	spectral ●	electronic ●	
data processing	calculation ●	computing ○	matching ●		
interpretation	selection ●	validation ●	deduction ●	prediction ●	
problem solving	identification ●	in/output ●	breakdown ○	methods ●	assembly ●
team work	skills identification ●	analysis ●	role assignment ○	organisation ○	resources ● outcomes ●
experiment design	input ●	output ●	precision ○	techniques ●	validity ●
communication	report ○	poster ○	oral ●	audience ●	
safety	COSHH ●	application ●	review ○	disposal ●	

Stereochemistry of nickel compounds

Four coordinate nickel complexes of the type $[NiX_2(PR_3)_2]$ (where $X = Cl^-$, Br^-, I^- or NCS^- and R = phenyl or cyclohexyl) can adopt either square planar or tetrahedral geometries. This team-focused activity involves synthesising a range of these complexes and determining the geometry as a solid and in solution by three different methods – the Evans NMR method, solid state magnetic moment and low frequency infrared.

Equipment

- pestle and mortar
- Quickfit condenser and tubing
- 100 cm^3 Quickfit round-bottomed flask
- 100 cm^3 Quickfit round-bottomed flask (two necked)
- 100 cm^3 heating mantle
- 25 cm^3 measuring cylinder
- 100 cm^3 measuring cylinder
- glass rod
- plastic bowl for ice
- sintered glass funnel and rubber seal
- 250 cm^3 Buchner flask attached to water pump via trap
- magnetic stirrer hotplate
- Schlenk tube
- sintered bubbler
- stirrer bar
- powder funnel

Reagents

- nickel(II) chloride.6H$_2$O
- nickel(II) bromide.3H$_2$O
- nickel(II) nitrate.6H$_2$O
- sodium thiocyanate
- sodium iodide
- triphenylphosphine
- tricyclohexylphosphine

■ carbon disulphide adduct

■ ethanol

■ absolute ethanol (dried)

■ propan-2-ol

■ propan-2-ol (dried)

Safety

Nickel salts: Carcinogenic.

Sodium thiocyanate: Harmful by ingestion, inhalation and skin contact. Irritating to eyes and skin.

Phosphines: Toxic by ingestion or skin absorption.

Carbon disulphide: Flammable. Toxic by inhalation, ingestion or skin absorption.

Alcohols: Flammable. Toxic by inhalation or ingestion.

Synthesis of nickel complexes

All the complexes are prepared by direct reaction of the appropriate nickel salt (chloride, bromide, iodide or thiocyanate) and phosphine. The thiocyanate salt of nickel is not available and must therefore be synthesised by a metathetical reaction in alcohol:

$$Ni(NO_3)_2 + 2NaNCS \rightleftharpoons Ni(NCS)_2 + 2NaNO_3$$

Sodium thiocyanate and sodium nitrate are relatively insoluble in alcohol and can be removed by filtration. Nickel iodide is made similarly, from nickel nitrate and sodium iodide.

Triphenylphosphine is reasonably air-stable but the corresponding tricyclohexyl compound is not and must be prepared by heating the air-stable carbon disulphide adduct under nitrogen.

$$PCy_3.CS_2 \rightleftharpoons PCy_3 + CS_2$$

All preparations of alkylphosphines of nickel should be carried out under nitrogen.

In all cases the nickel phosphine complexes are made by adding a solution of the nickel salt to a stirred and refluxing solution of the phosphine. Use the quantities indicated in Table 1. Before starting the synthesis discuss the procedure and apparatus with a demonstrator.

Table 1

(a) Triphenylphosphine complexes

Complex	PPh$_3$ /g	solvent /cm^3	nickel salt /g	solvent* /cm^3
[NiCl$_2$(PPh$_3$)$_2$]	2.8	30 (dried absolute ethanol)	NiCl$_2$.6H$_2$O 1.2 g	15
[NiBr$_2$(PPh$_3$)$_2$]	2.8	30 (propan-2-ol)	NiBr$_2$.3H$_2$O 1.4 g	15
[NiI$_2$(PPh$_3$)$_2$]	2.8	30 (dried absolute ethanol)	NiI$_2$ from 2.0 g NaI + 1.5 g Ni(NO$_3$)$_2$.6H$_2$O	15
[Ni(NCS)$_2$(PPh$_3$)$_2$]	2.8	25 (propan-2-ol)	Ni(NCS)$_2$ from 1.5 g Ni(NO$_3$)$_2$.6H$_2$O + 0.8 g NaNCS	15

*dried, absolute ethanol is used in all cases.

(b) Tricyclohexylphosphine complexes

Complex	PCy$_3$.CS$_2$ /g	solvent* /cm^3	nickel salt /g	solvent* /cm^3
[NiCl$_2$(PCy$_3$)$_2$]	1.9	20	NiCl$_2$.6H$_2$O – 0.6 g	15
[NiBr$_2$(PCy$_3$)$_2$]	1.9	20	NiBr$_2$.3H$_2$O – 0.7g	15
[NiI$_2$(PCy$_3$)$_2$]	3.8	40	NiI$_2$ from 2.0 g NaI + 1.5 g Ni(NO$_3$)$_2$.6H$_2$O	15
[Ni(NCS)$_2$(PCy$_3$)$_2$]	3.8	15	Ni(NCS)$_2$ from 0.7 g Ni(NO$_3$)$_2$.6H$_2$O + 0.4 g NaNCS	8

*dried, absolute ethanol is used in all cases.

Structure determination

Three different methods are used to distinguish between the four-coordinate square planar and tetrahedral geometries. Two of these involve a magnetic moment determination (in solution by the Evans NMR method and in the solid using a magnetic balance) and the third uses the nickel halogen bond stretch in the far infrared.

At this stage discuss these methods with a demonstrator who will provide instruction as to the quantity of materials to use and how to operate the instruments.

D21

Stereochemistry of nickel compounds

Four coordinate nickel complexes of the type $[NiX_2(PR_3)_2]$ (where X = Cl^-, Br^-, I^- or NCS^- and R = phenyl or cyclohexyl) can adopt either square planar or tetrahedral geometries. This team-focused activity involves synthesising a range of these complexes and determining their geometry as solids and in solution by three different methods – the Evans NMR method, solid state magnetic moment and low frequency infrared. Students identify the factors influential in determining the geometry and are asked whether it is possible to design a complex that exhibits both geometries. The strategy, experimental design and subsequent work involve extensive team discussions and practical work.

When students have completed the activity, they should be able to:

■ identify the features and equipment appropriate to advanced chemical laboratories;

■ work safely, and be familiar with the safety equipment, precautions and procedures for each experiment performed, including any new experiments which must be individually COSHH assessed;

■ operate equipment and follow experimental procedures correctly. These may include: reactions in an inert atmosphere, magnetic moment measurements, recording of infrared and NMR spectra;

■ perform a chemical experiment following published details;

■ maintain an accurate record of work so that others are able to understand the results and repeat the experiments;

■ interpret infrared and NMR spectra in the context of these experiments;

■ correlate the experimental findings made by their laboratory group with their theoretical studies; and

■ develop extensions to the work which would further investigate factors affecting the geometry of four-coordinate metal complexes.

The work takes between four and six laboratory sessions, depending upon how far the project is pursued and how many complexes each student makes.

Pre-laboratory work

It cannot be emphasised too strongly that the pre-laboratory session and the intermediate briefing/discussion are vital for the success of the project. The pre-laboratory session should involve a student-active discussion of the following:

■ geometric possibilities for four-coordinate complexes of nickel;

■ d orbital electronic configurations for those geometries; and

■ the consequences of the configurations in terms of the magnetic moment.

Additionally, indicate how the Ni-X stretch (X = Cl, Br, I) region 500–190 cm^{-1} can be used to distinguish the trans-planar geometry from the tetrahedral or cis-planar geometry.

Safety precautions should also be outlined in the pre-laboratory session. Note that all reactions involving tricyclohexylphosphine should be carried out under nitrogen.

Laboratory session

The first two laboratory sessions should be used for the synthesis and structure determination of the complexes. The ideal set up is to divide students into groups of four pairs so that each group synthesises one sample of each of the eight nickel complexes, [NiX$_2$(PR$_3$)$_2$] (where X = Cl$^-$, Br$^-$, I$^-$ or NCS$^-$ and R = phenyl or cyclohexyl). No details for the use of the magnetic balance, NMR Evans method or infrared spectrophotometer are given here as instructions will be specific to the available instruments.

The two cyanato complexes are planar, the remaining triphenylphosphine complexes are tetrahedral and the remaining tricyclohexyl complexes are planar. The discussion should centre on the fact that electronic factors are as influential as steric factors.

Further Investigations

Is it possible to design a complex that exists in both planar and tetrahedral forms? There is a range of possible new phosphines to use and it is valuable to get the students to make suggestions. It might be possible to make two further sets of four complexes; one set with dicyclohexylphenylphosphine and one with cyclohexyldiphenylphosphine. These phosphines can be synthesised via a Grignard reaction using bromocyclohexane and either dichlorophenylphosphine or chlorodiphenylphosphine. Other useful ligands include mixed phosphines involving 2-methoxyphenyl, 3-methoxyphenyl, 4-methoxyphenyl, 2-methylphenyl, 3-methylphenyl, 4-methylphenyl, cyclohexenyl groups and even bidentate phosphines such as bis(dimethylphenylphosphino)ethane.

If a Grignard (or other) synthesis of ligands is required, then two more (total of four) laboratory sessions will be required to complete the project.

T21

Stereochemistry of nickel compounds

Equipment

- pestle and mortar
- Quickfit condenser and tubing
- 100 cm^3 Quickfit round-bottomed flask
- 100 cm^3 Quickfit round-bottomed flask (two necked)
- 100 cm^3 heating mantle
- 25 cm^3 measuring cylinder
- 100 cm^3 measuring cylinder
- glass rod
- plastic bowl for ice
- sintered glass funnel and rubber seal
- 250 cm^3 Buchner flask attached to water pump via trap
- magnetic stirrer hotplate
- Schlenk tube
- sintered bubbler
- stirrer bar
- powder funnel

Reagents

- nickel(II) chloride.$6H_2O$
- nickel(II) bromide.$3H_2O$
- nickel(II) nitrate.$6H_2O$
- sodium thiocyanate
- sodium iodide
- triphenylphosphine
- tricyclohexylphosphine
- carbon disulphide adduct
- ethanol
- absolute ethanol (dried)
- propan-2-ol
- propan-2-ol (dried)

Access to a nitrogen Schlenk line is required.

Additional chemicals and equipment will be required if the projects involve the synthesis of further phosphine ligands.

Data will be collected using an NMR (Evans method), infrared (500–190 cm^{-1}) and a magnetic balance.

...BRAVEN...
...ARTS LTD
TITHEBARN STREET
LIVERPOOL L2 2ER
TEL. 0151 231 4022